Dedicated to
my son
Adam

Individually and collectively, the Fellows of the Royal Society were involved in many wartime activities. For example, prior to the outbreak of World War Two Her Majesty's Government decided to establish a Central Register of persons with 'professional, scientific, technical or higher administrative qualifications', for use in time of War, and entrusted this work to the Ministry of Labour. The Secretaries of the Royal Society discussed the project with officials from that Ministry, and at the beginning of 1939, started to compile the part of the Central Register which dealt with scientific research. Some Fellows such as Sir Winston Churchill FRS, Albert Einstein ForMemRS, Alan Turing FRS, Sir Barnes Wallis FRS and Jan Christian Smuts FRS were either in the public eye during World War Two, or were the subject of films and documentaries soon afterwards. Equally there are individuals who have escaped the limelight. For example, Sir Harry Work Melville FRS was Scientific Adviser to the Chief Superintendent of Chemical Defence, Ministry of Supply (1940–1943) based mainly at Porton Down and Superintendent of the Radar Research Station at Malvern (1943–1945). Sir (Thomas) Angus Lyall Paton FRS organised staff for the supervision of a number of the reinforced concrete caissons and Phoenix units that formed part of Mulberry Harbours for the invasion of France, which incidentally were designed by Sir Bruce White. Of course Sir Frank Whittle OM CBE CB FRS also came to the public's attention for his war work on the jet engine. Interestingly, William Michael Herbert Greaves FRS was appointed Astronomer Royal for Scotland, as well as Professor of Astronomy at Edinburgh University and was in charge of the Royal Observatory in Edinburgh. He helped to set up an independent time service there, in case the regular Greenwich time service should be completely disrupted. This service was run by Greaves with the help of only a very small staff. These, and many other stories, provide a fascinating and detailed picture of the men frequently labelled 'boffins', and the work they did during World War Two.

Dave Rogers set up Danercon Ltd in 2004, having previously worked for a multinational company for 23 years. During his industrial career, Dave spent time working in research and development and in the manufacturing division. His research experience involved product component research, product design and the implementation of process verification equipment. Dave's manufacturing experience covers the product issues of day to day manufacture, product design as part of a waste reduction effort, as well as leading a process research and development group of some twenty engineers and scientists.

Dave holds Bachelor and Doctorate degrees in Chemistry, Fellowships with the Royal Society of Chemistry, The Royal Photographic Society and the British Institute of Professional Photography (the latter by invitation). He was Visiting Professor at the University of Westminster 2002–2005. Dave is a long term School Governor having recently completed fifteen years as Primary School Governor.

Dave has written or edited nine books. Two are war related which he edited for his father, a third wartime book was co-written by Dave and his father. This is Dave's second book for Helion, having previously written *Top Secret: British Boffins in World War One*.

Men Amidst the Madness

British Technology Development in World War II

David Rogers

Helion & Company

Helion & Company Limited
26 Willow Road
Solihull
West Midlands
B91 1UE
England
Tel. 0121 705 3393
Fax 0121 711 4075
email: info@helion.co.uk
website: www.helion.co.uk

Published by Helion & Company 2014
Designed and typeset by Bookcraft Ltd, Stroud, Gloucestershire
Cover designed by Euan Carter, Leicester (www.euancarter.com)
Printed by Lightning Source, Milton Keynes, Buckinghamshire

ISBN 978-1-909982-08-6

British Library Cataloguing-in-Publication Data
A catalogue record for this book is available from the British Library

For details of other military history titles published by Helion & Company Limited
contact the above address, or visit our website: http://www.helion.co.uk

We always welcome receiving book proposals from prospective authors working in
military history.

Contents

List of Figures

Preface

The Royal Society Fellows have knowledge and skills in many fields of engineering, medicine, science and technology. Within their own fields they are high profile individuals yet it is doubtful that there will ever be a definitive list of who did what during World War Two. Any snapshot such as this will always exclude some Fellows. For example most of the data presented below is of Fellows or Foreign Members who have died and for which biographical details are available. Even if the author asked the Royal Society to contact living Fellows, there would still be a gap of those who are sadly no longer with us, but for whom no biographical details have yet been published.

One must therefore accept that the material below is typical of activities of Fellows of the Royal Society and is not exhaustive.

Individually and collectively, the Fellows were involved in many wartime activities. For example, prior to the outbreak of World War Two Her Majesty's Government decided to establish a Central Register of persons with 'professional, scientific, technical or higher administrative qualifications', for use in time of War, and entrusted this work to the Ministry of Labour. The Secretaries of the Royal Society discussed the project with officials from the Ministry of Labour, and at the beginning of 1939, started to compile that part of the Central Register which dealt with scientific research.

An Advisory Council for the Central Register was set up by the Ministry of Labour under the chairmanship of Sir Walter Moberley, by which various Committees were established including the Scientific Research Committee which had for its Chairman Professor A.V. Hill FRS. Sub-Committees were then formed by this Scientific Research Committee to deal with the ten classified groups of bacteriology (human), botany, chemistry, engineering sciences, geology/mineralogy, mathematics, pathology, physics, physiology, psychology and zoology in various branches of science. The members of these Sub-Committees were drawn up mainly from the Fellowship of the Society and acted as advisory panels to the Ministry of Labour in the task of selection. By the middle of October 1939 a total of 6,484 individuals had been registered through the Society.

Although this activity was widely known within the Royal Society other activities were kept relatively secret until 1946 when some details were published.

Some Fellows such as Sir Winston Churchill FRS, Albert Einstein ForMemRS, Alan Turing FRS, Sir Barnes Wallis FRS and Jan Christian Smuts FRS were either in the public eye during World War Two, or were the subject of films and documentaries afterwards. Equally there are individuals who have escaped the limelight.

For example, Sir Harry Work Melville FRS was Scientific Adviser to the Chief Superintendent of Chemical Defence, Ministry of Supply (1940–1943) based mainly at Porton Down and Superintendent of the Radar Research Station at Malvern (1943–1945). Sir (Thomas) Angus Lyall Paton FRS organised staff for the supervision of a number of the reinforced concrete caissons and Phoenix units that were part of Mulberry Harbour for the invasion of France, which incidentally were designed by Sir Bruce White.

Interestingly, William Michael Herbert Greaves FRS was appointed Astronomer Royal for Scotland, as well as Professor of Astronomy at Edinburgh University and was in charge of the Royal Observatory in Edinburgh. He helped to set up an independent time service in Edinburgh, in case the regular Greenwich Time service should be completely disrupted. This service was run by Greaves with the help of only a very small staff. There are many other stories, just a few of which appear below.

Acknowledgements

The opportunity to work for a Fellow of the Royal Society during your education should not be passed up lightly, if at all. I had the opportunity during the late 1970s and have never regretted it. My particular Fellow remains a high profile organic chemist in 2013, even though he is in his 80s. Looking back on my experience it is not just the hard work and dedication, some of which inevitably rubbed off on me, but also some of the life skills one learns along the way.

For example every man, woman and child on the planet only has 24 hours in the day. The Fellows have developed their organisation skills to such a degree that they are able to maximise their time to be the most productive possible. Of course they delegate some of the day to day tasks to post-doctoral demonstrators and members of their own staff, however in delegating some tasks they always have a knack of keeping abreast of daily events, even if they are on some lecture tour. Of course they have friends and contacts who are also Fellows who visit from time to time, presenting those involved with visits an opportunity to interact with the best available discussions on a variety of topics.

I cannot speak of Fellows from history, however there is nothing to suggest that this situation has changed much since the Royal Society was founded in the early 1660s.

In writing this volume of the involvement of wartime Fellows in the machinery of War, I was once more presented with the problem of how to address those mentioned below. By that I mean the form of address, such as Sir or Lord. It is an inevitable consequence of their activities that some were knighted and a few ennobled. It makes life difficult for a biographer – especially if during the process the individual changes names completely as was the case for Frederick Alexander Lindemann FRS. Elected to Fellowship in 1920, Lindemann was ennobled in 1942 as Baron Cherwell, and raised further to Viscount in 1956.

There is also the issue of those individuals elected to Fellowship during the War. Does one ignore their contribution until election or present all of their war-work. A decision was needed which added consistency within this book and also with the earlier volume dealing with the contributions of Fellows to World War One. I have maintained my position and decided to present individuals with the end of career titles and honours. This also includes Winston Churchill, who was not knighted until 1953.

Finally those surrounding me during this effort need to be recognised for their patience in listening to me for endless hours. Particularly, my wife Carolyn and sons Adam and James. I also would like to take the opportunity to recognise the discussions with Bill Broadhurst. Always a delight to talk with, Bill made some comments worthy of inclusion. Thank you to one and all. Of course the Library Staff of the Royal Society helped with timely discussions, retrieval of documents, photographs of Fellows etc. Without their help this book would be less complete or dare I say it not exist at all! Thank you.

1 Introduction, administration and displaced persons

The Royal Society is the oldest scientific society in the world. Founded on 28 November 1660 with a Royal Charter from King Charles II in 1661, it has recruited to its Fellowship (or Foreign Membership) most of the outstanding scientists and engineers since that time. From its humble accommodation in Gresham College to its present 'home' in Charlton House Terrace (the previous accommodation being Burlington House), the Royal Society has continued to promote excellence in science. To that end the Society publishes several in-house scientific journals, and internal pamphlets and books including *Notes and Records* which provide a snapshot of events for a given year and a Biographical Memoir of most Fellows and Foreign Members, written by their peers. For example, there is a short reference in *Notes and Records*, which states:

> The offices of the Society with the exception of two departments returned to Burlington House from their temporary evacuation quarters at Trinity College Cambridge on 20th March 1940. The Central Register and the Departments are occupying premises at Meadway Gate Golders Green NW1.[1]

Founded by King Henry VIII in 1546, Trinity College (see figure 1.1), has enjoyed a long associated with the Royal Society. Indeed of the 160 or so current College Fellows two are former Presidents of the Royal Society (Sir Michael Atiyah OM FRS and Lord Rees of Ludlow OM FRS). In addition to these former Royal Society Presidents there are many other Fellows.

The external frontage provides but a glimpse of the large number of rooms and rich history.

The entry in *Notes and Records* makes no comment on the reason for the departure to Cambridge nor for the return to Burlington House. In many respects, Burlington House was pivotal as a general meeting place during World War Two, at least for the scientific community. It is just a short walk from there to St James' Square and thence to various Government buildings, and an even shorter walk for the President of the Royal Society, at least in the early days of the War.

Sir William Henry Bragg OM FRS was elected President in 1935 and served for a five year term until 1940. Sir William was an active scientist in World War One. He was Resident Director of Research at the Admiralty Experimental Station (1916–1917) and Scientific Member of the Anti-submarine Division in the Admiralty (1918) to name just two of his wartime activities. Incidentally, he was also an Honorary Fellow of Trinity College, Cambridge having been elected in 1920. Of interest here was his World War Two 'day job' as Director of the Royal Institution, Fullerian Professor of Chemistry at the Royal Institution and Director of Davy-Faraday Research Laboratory from 1923.

Figure 1.1 Trinity College Cambridge. (Author's photograph)

Burlington House, then the home of the Royal Society, is located on Piccadilly in London and the Royal Institution on Albemarle Street off Piccadilly. It is likely that Sir William enjoyed a short walk from his apartment in the Royal Institution to the rooms of the Royal Society, spending most of his active War years living and working in a small area of Central London.

Of course there were many other scientists active in World War One, who also served in a senior capacity in World War Two. Figure 1.3 shows the Presidents of the Royal Society during and following World War Two. It is shown here as the names on this list were all active in many diverse capacities, and mentioned on numerous occasions in later chapters. Incidentally, these individuals were also Nobel Laureates and elected to the Order of Merit.

Figure 1.2 Burlington House. (Author's photograph)

Name	President for the years	Nobel Prize year	Scientific discipline
Bragg, Sir William Henry	1935–1940	1915	Physics
Dale, Sir Henry (Hallett)	1940–1945	1936	Physiology or Medicine
Robinson, Sir Robert	1945–1950	1947	Chemistry

Figure 1.3 Some Presidents of the Royal Society.

Of course the President works with senior Officers of the Society and indeed a Council helping to shape scientific direction. As with the five year term of President, the War years saw some stability in the four senior Officers of Treasurer, Biological Sciences Secretary ('Sec. A'), Physical Sciences Secretary ('Sec. B') and Foreign Secretary. These individually were:

Sir Thomas Ralph Merton FRS (1939–1956); Archibald Vivian Hill FRS (1935–1945); Sir Alfred Charles Glyn Egerton FRS (1938–1948); Sir Henry Thomas Tizard FRS (1940–1945) respectively.[2]

All of the Officers mentioned above and the Presidents in figure 1.3 provided unstinting service to the Government and its Departments during the War (Sir William sadly died on 12 March 1942). Fortunately they were not unique amongst the Fellowship for offering their time for the war effort. Chapter 2 outlines the Royal Society War Committees covering many aspects of the war effort. Additionally there were many Fellows who acted independently giving their time and resources to the War. Some Fellows also worked in Government, more of which below.

Arguably, some Fellows defy categorisation. Perhaps the most unusual activity might be that of William Michael Herbert Greaves FRS, see figure 1.4. Karen Moran of The Royal Observatory, Edinburgh kindly provided the following information:

> By 1943 the work of the Royal observatory was reduced to routine activities such as the local time service and the maintenance of meteorological and seismological records. However, a new and important program of work had appeared ... in the autumn of 1940 when the country became the target of severe air raids Sir Harold Spencer Jones suggested the [ROE] ought to be equipped for the provision of an independent national time service in case the normal service from Greenwich's station at Arbinger were put out of action. Greaves took up this major new task with enthusiasm and managed with the sole assistance of two members of Greenwich staff to complete the installation of all necessary equipment by the end of the year and to operate the Rugby rhythmic time signals in January 1941 when the Greenwich time-service was in fact disrupted by enemy action.

Greaves kept this national time service going throughout the whole War checking the standard clocks whenever possible by transit observations. As a by-product of this work he made a careful comparison of the performance of the Edinburgh and Greenwich free pendulum Shortt clocks with that of quartz crystal clocks newly installed by the Post Office. The result, showing the definite superiority of quartz over even the very best pendulum clocks was published, by Greaves and Symms, in 1943 in the MNRAS. The same journal carried in 1945 and 1946 Greaves' accounts of Edinburgh's contribution to the wartime national service which was suspended only in Feb 1946.[3]

Figure 1.4 William Michael Herbert Greaves. (© Godfrey Argent Studio. Reproduced with permission)

Interestingly, various Government figures were Fellows. The 1941 *Whitaker's Almanac* lists the following:

Government Offices held by Fellows 1941

National War Cabinet:

Sir Winston Churchill FRS	Prime Minister
Sir John Anderson FRS	Lord President of the Council
Clement Atlee	FRS Lord Privy Seal
Viscount Halifax	Secretary of State for Foreign Affairs
Arthur Greenwood	Minister without Portfolio
Sir Kingsley Wood	Chancellor of the Exchequer
Ernest Bevin	Minister of Labour and National Service

The above formed the War Cabinet. The following were in the National War Cabinet in addition to those mentioned above:

Albert Victor Alexander	First Lord of the Admiralty
Anthony Eden	Secretary of State for War
Sir Archibald Sinclair	Secretary of State for Air
Viscount Simon	Lord Chancellor
Herbert Morrison	Home Secretary and Minister for Home Security
Viscount Cranborne	Secretary of State for the Dominions
Lord Lloyd	Secretary of State for the Colonies
Captain Oliver Lyttelton	President of the Board of Trade
Sir Andrew Rae Duncan	Minister of Supply
Alfred Duff-Cooper	Minister of Information
Leopold Stennett Amery	Secretary of State for India and Burma
Malcolm MacDonald	Minister of Health
Lord Woolton	Minister of Food
Ernest Brown	Secretary of State for Scotland
Herwald Ramsbotham	President of the Board of Education
Robert Hudson	Minister of Agriculture
J.T.C. Moore-Brabazon	Minister of Transport
Ronald Hibbert Cross	Minister of Shipping
Hugh Dalton	Minister of Economic Warfare
Lord Hankey FRS	Chancellor of the Duchy of Lancaster
Sir W.J. Womersley	Minister of Pensions
Lord Reith	Minister of Works and Buildings and First Commissioner of Works
W.S. Morrison	Postmaster-General

Sir Donald Somervell	Attorney-General
Sir William A. Jowitt	Solicitor-General
Thomas Cooper	Lord Advocate
J.S.C. Reid	Solicitor-General for Scotland[4]

Some Fellows were not in the Government, yet were directed by Cabinet Ministers. Some were appointed to the Cabinet after 1941 and so do not appear on the list above. For example:

- Sir Richard Stafford Cripps FRS 'was asked by Churchill to go as British Ambassador to Moscow ... on his return Churchill asked him to become Leader of the House of Commons with the office of Lord Privy Seal and a place on the War Cabinet ... Churchill also asked him to become Minister of Aircraft Production without a place on the War Cabinet.'[5]
- Jan Christian Smuts FRS 'subsequently went to London to serve in the War Cabinet as a Minister without Portfolio ... he was again in constant consultation with universal benefit to the Allied cause throughout World War Two.'[6]
- Reginald Edward Stradling FRS 'formed the Air Raid Committee. This Committee, when it began its work, found that no detailed information about the history of an explosion such as that if a large bomb was available, and their obvious first step was to fill this serious gap in essential knowledge.'[7]
- Hendrick Johannes van der Bijl FRS 'was born in Pretoria on 23rd November 1887. In November 1939, General Smuts, who was both Prime Minister and Minister of Defence, set up what was virtually a Ministry of Supply with van der Bijl at its head.'[8]
- Frederick Alexander Lindemann FRS 'was Sir Winston Churchill's Personal Assistant, a post he continued to hold when Churchill became Prime Minister, until he became a formal member of the Government as Paymaster General at the end of 1943, by which time he had been raised to the peerage with the title Baron Cherwell of Oxford.'[9]
- Edward Ettingdean Bridges FRS 'took up the appointment as Secretary of the Cabinet and Permanent Secretary of the combined offices of Cabinet (1st August 1938), Committee of Imperial Defence, Economic Advisory Council and Minister for the Coordination of Defence.'[10]
- Stanley Melbourne Bruce FRS 'Privy Councillor, former Prime Minister of Australia and High Commissioner for Australia in London (in 1944).'[11]

In addition to the Fellows in Government, there were also many who occupied positions in Government organisations. The information in Appendix I details the contents of two letters, one written in August 1942 and the other a month later. They are both addressed to an individual called Mr Lucker and are from Mr H. Everett who was at the time one of the Secretaries of both the Scientific Advisory Committee and of the Engineering Advisory Committee. Unfortunately the letter from Mr Lucker to Mr Everett has not been filed with the replies in the National Archives folder. One can only assume that there was an attempt to document the various Government Committees with, in some cases, their full membership.

Figure 1.5 was also included in with the two letters, proving as useful now as it must have been at the time.

Even though Appendix I and figure 1.5 show the range of some of the Government Organisations, they are not exhaustive. *Whitaker's Almanack* (1941) also published some of the other Committees and Government Departments. Some of that information is presented below for completeness.

Royal Botanic Gardens, Kew

Sir A.W. Hill FRS	Director and Technical Advisor in Botany
J. Hutchinson FRS	Keeper of Museums
W.B. Turrill FRS	Botanist

Executive Council of the Imperial Agricultural Bureaux

F.J. du Toit	Chair
Sir David Chadwick	Secretary
Sir Edward Russell FRS	Director, Rothamsted Experimental Station
W. Horner Andrews	Director, Veterinary Research Laboratory, Weybridge
Baron Boyd-Orr FRS	Director, Rowlett Research Institute, Aberdeen
Professor F.A.E. Crew FRS	Director, Institute of Animal Genetics, Edinburgh
Sir Frank Engledow FRS	Director, Plant Breeding Institute, Cambridge
Sir Reginald Stapledon FRS	Director, Welsh Plant Breeding Station, Aberystwyth
Sir Ronald Hatton FRS	Director, East Malling Research Station, Kent
Robert Leiper FRS	Director, Institute of Agricultural Parasitology, St. Albans
Professor H.G. Champion	Director, Imperial Forestry Institute, Oxford
Professor Herbert Kay	Director, National Institute for Research in Dairying, Shinfield
Sir Guy Marshall FRS	Director, Imperial Bureau of Entomology, Natural History Museum

S.P. Wiltshire   Director, Imperial Mycological Institute, Kew

Ministry of Aircraft Production
Lord Beaverbrook   Minister of Aircraft Production
A.S. Quartermain   Director-General of Aircraft Production

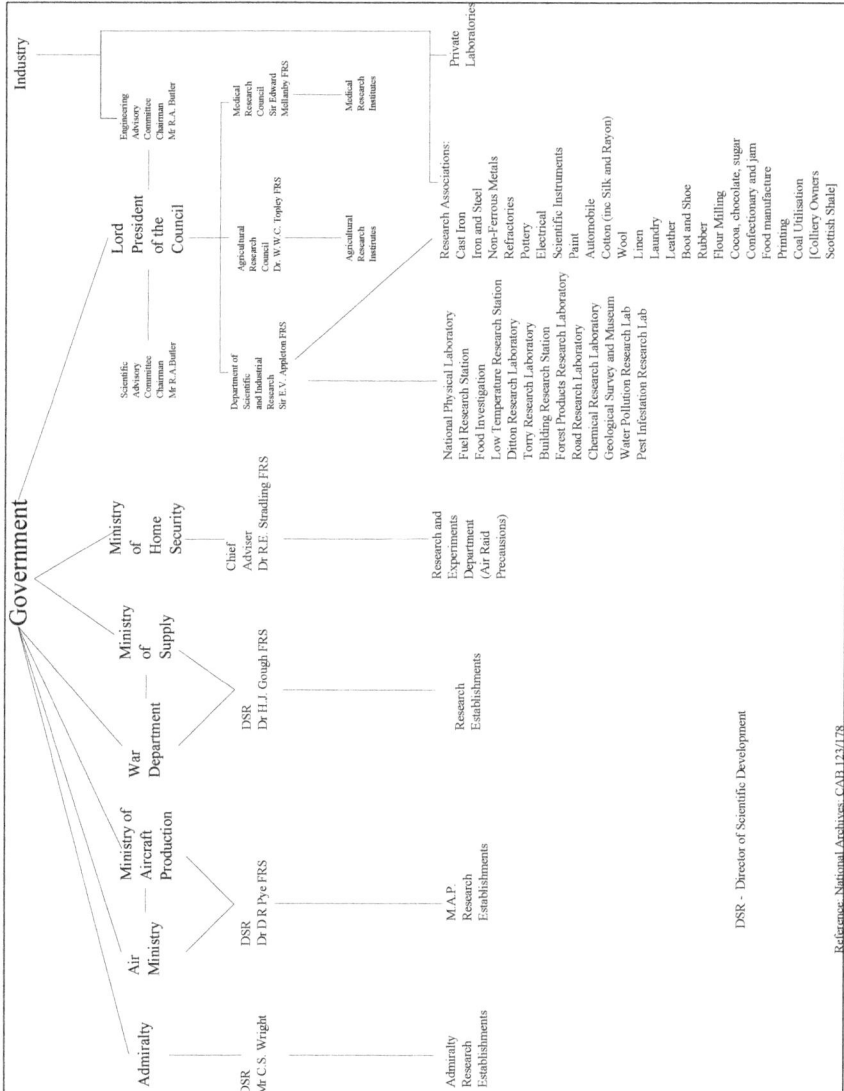

Figure 1.5 Government Research Organisations.

	Factories
W.L. Stephenson	Director-General of Equipment
Arthur R. Cooper	Director of Air Ministry Factories
T.C.L. Westbrook	Controller of Purchases
Sir Frank Edward Smith FRS	Controller of Telecommunications Equipment[12]

The Government Chemist

Sir John Jacob Fox FRS Government Chemist[13]

The British Museum

Standing Committee

The Archbishop of Canterbury	Principal Trustee
The Lord High Chancellor	Principal Trustee
The Speaker of the House of Commons	Principal Trustee
The Earl of Harewood	ex officio – appointed by the Sovereign

Earl of Crawford
Earl of Ilchester
Earl Baldwin of Bewdley FRS
Lord Harlech
Lord Macmillan
Sir Ronald William Graham
Sir Wilfred Greene
Sir Henry Tizard FRS
Sir William Bragg PRS
Sir C. R. Peers
Sir Albert Charles Seward FRS
George M. Trevelyan FRS
F. Cavendish Bentick
Professor G. A. Murray
Professor John Stanley Gardiner FRS
C. H. St. John Hornby[14]

The British Museum (Natural History)

Sir Clive Forster-Cooper FRS	Director
M.A.C. Hinton FRS	Keeper of Zoology
N.D. Riley	Keeper of Entomology
W.N. Edwards	Keeper of Geology
W. Campbell Smith	Keeper of Mineralogy
J. Ramsbottom	Keeper of Botany[15]

Royal Observatories

Greenwich	Sir Harold Spencer-Jones FRS
	Astronomer Royal
Cape of Good Hope	John Jackson FRS Astronomer
Edinburgh	William Greaves FRS Astronomer
	Royal for Scotland[16]

Geological Survey of Great Britain and Museum of Practical Geology
Directory of Survey and Museum E.B. Bailey FRS

Unlike no other War before it, there were unsettling times across Europe from 1933, arguably since the election of Adolf Hitler to the post of Chancellor. There is a brief account of the period 1933 to 1939 below as this period led to an influx of scientists and engineers into Britain and indeed the United States. Included in the list of those which fled the Nazi regime are some household names.

Before that, however, it is worth briefly recording the fate of two Fellows born closer to home.

Born on 28 Jan 1903 in Newbridge Southern Ireland, Dame Kathleen Lonsdale FRS was one of the first two female scientists to be elected Fellow of the Royal Society (the other being Marjory Stephenson). Her father, Fred, was a wireless operator and received the last signals from the Titanic in 1912.

Unusually for a young lady at that time, Lonsdale read physics at Bedford College where she came into contact with Sir William Bragg FRS. At his request she joined his group moving to the Royal Institution when Bragg moved there in 1923. Later, Lonsdale was sent to Holloway Prison for a month for failing to pay the £2 fine for not registering for employment or Civil Defence duties. In her defence she had three young children at the time.[17]

Interestingly, Sir Henry Dale sent her scientific papers and instruments in prison so she would not get bored!

Samuel Victor Perry FRS was born on 16 July 1918. Perry's Royal Society Biographical Memoir records the following:

> His real adventures seem to have followed capture by the Germans in 1942. His first escape was from Bologna in Italy in 1943, only for him to be caught soon afterwards and moved to Germany. On the way he escaped again this time to Mantua, hoping to get to Switzerland. By coincidence he was captured crossing a bridge by a member of the same unit that had caught him earlier as it too had moved to Bologna. The second escape within such a short time aroused the respect of the officers commanding the unit and they offered Perry wine at their table which he gratefully accepted, before they locked him up again! After experiencing three more prisons in Germany

Figure 1.6 Kathleen Lonsdale. (© Godfrey Argent Studio. Reproduced with permission)

he was moved to Moravská Třebová now in the Czech Republic. In 1944 he managed to escape whilst being transported on a train by cutting a hole in the wagon as it moved through Bavaria. His escape was again short lived. Prisoners were not punished for escaping, however he had damaged a railway carriage. During the day he went for his trial in Hildesheim, the Brunswick camp where he was imprisoned was bombed in an air raid and several of his fellow prisoners were killed.[18]

Figure 1.7 Samuel Victor Perry. (© Godfrey Argent Studio. Reproduced with permission)

Displaced Persons

The Council for Assisting Refugee Academics (CARA) was founded in 1933 as the Academic Assistance Council (AAC), changing its name in 1936 to Society for the Protection of Science and Learning prior to the final name change to CARA. The CARA website posts the following account of its history:

> The Academic Assistance Council (AAC) was established in April 1933 by William Beveridge who enlisted the help of Lord Ernest Rutherford FRS and A.V. Hill FRS. Some of the most influential scientists of the day signed their founding statement including:
>
> J.B.S. Haldane FRS
> Lord Rayleigh FRS
> Sir Charles Scott Sherrington FRS
> Sir Frederick Gowland Hopkins FRS
> Sir Joseph John Thomson FRS
> Sir Robert Robinson FRS
> Sir William Bragg FRS

Ten thousand people attended an AAC event at the Royal Albert Hall in London at which Professor Albert Einstein ForMemRS (Foreign Member of the Royal Society) spoke. In his address Einstein commented (full speech available through the CARA website):

> I am glad that you have given me the opportunity of expressing to you here a deep sense of gratitude as a man, as a good European, and as a Jew ... It cannot be my task today to act as judge of the conduct of a nation which has for many years considered me as her own; it is an idle task to judge in times when action counts. Today, the questions concerns us are: how can we save mankind in its spiritual acquisitions all which we are the heirs? How can one save Europe from a new disaster?[19]

By the start of World War Two, the organisation had saved 900 academics from Nazi persecution. Some of the records of the organisation for this time frame can be found online others in the Bodleian Library in Oxford.[20]

The CARA website list many internationally renowned scientists it has assisted over the years. Figure 1.8 shows just those individuals who were awarded the Nobel Prize. Incidentally not all of these scientists were Fellows or Foreign Members of the Royal Society.

Name	Nobel Prize
Bethe, Hans Albrecht	Physics 1967
Born, Max	Physics 1954
Chain, Sir Ernst	Physiology or Medicine 1945
Delbruck, Max	Physiology or Medicine 1969
Gabor, Dennis	Physics 1971
Herzberg, Gerhard	Chemistry 1971
Heyrovsky, Jaroslav	Chemistry 1959
Katz, Sir Bernard	Physiology or Medicine 1970
Krebs, Sir Hans	Physiology or Medicine 1953
Lipmann, Fritz	Physiology or Medicine 1953
Loewi, Otto	Physiology or Medicine 1936
Luria, Salvador E.	Physiology or Medicine 1969
Ochoa, Severo	Physiology or Medicine 1959
Perutz, Max	Chemistry 1962
Polanyi, John C	Chemistry 1986
Segrè, Emilio Gino	Physics 1959

Figure 1.8 Future Nobel Laureates helped by CARA.

The CARA website also mentions that in total some 1,500 academics have been assisted over the years, eighteen received knighthoods and over 100 were elected Fellows or Foreign Members of the Royal Society and/or the British Academy. In reviewing the careers of the individuals mentioned in figure 1.8, and some of the Fellows listed below, it is worth remembering that it is not just their wartime exploits that are of interest. Their whole careers outside of their homelands were distinguished.

The following are just by way of example and are documented here to show the contributions from some of these individuals.

Max Perutz FRS worked in the Cambridge University Cavendish Laboratory in 1937 both as a research student (under J.D. Bernal FRS) and as an assistant to Sir Lawrence Bragg FRS. He was interned as an 'enemy alien' but later released to work for the military, see below. His post War career included: ICI Fellow (1945–1947); head of MRC Unit for the Study of Molecular Structure of Biological Systems (1947–1957); Reader in Chemistry, Davy Faraday Research Laboratory, Royal Institution, London (1954–1958); Director, MRC Laboratory of Molecular Biology, Cambridge (1957–1979,), Chairman (1967); Honorary Fellow, Peterhouse, Cambridge (1962); Chairman, European Molecular Biology Organisation (1963–1969); Fullerian Professor of Physiology, Davy Faraday Laboratory, Royal Institution (1973–1979); honours and awards: Nobel Prize (jointly with Sir John Kendrew FRS) 1960, CBE (1963), CH (1975), OM (1988), Millennium FRSC (2000); Lectures: Croonian 1968; Humphry Davy 1988; Medawar 1992

Once released from internment in Canada, the return journey being tortuous and lengthy, Perutz was recruited by Bernal (his former supervisor), to work on Project Habakkuk. This project was so secret at the time that its leader Geoffrey Pyke reportedly commented that only Bernal, Lord Mountbatten and himself knew about it. The basic concept was to strengthen ice floes with composite materials such as wood pulp or cotton wool, such that they could be used as floating airfields in the Atlantic Ocean. Perutz's Royal Society Biographical Memoir comments:

> assisted by a detail of commandos he produced a composite of ice and wood pulp as strong, weight for weight, as concrete, and resistant to rifle bullets, which they named Pykrete after their leader. In a demonstration called by Mountbatten to the Joint Chiefs of Staff (from which Max was absent) a revolver was fired into a block of ice, which duly shattered. When a similar shot was fired into 'pykrete', the bullet rebounded and hit the Chief of the Imperial General Staff on the shoulder. The Chief was unhurt, but Habakkuk was under a cloud.[21]

The following is a partial list of Fellows or Foreign Members who suffered internment during the War:

- Albrecht Fröhlich FRS 'worked as an Artisan Class I (electrical), from April 1942 until the end of October 1945. For a while he was interned, presumably as an enemy alien.'[22]
- André Weil ForMemRS 'was in Finland when War was declared and was nearly shot as a spy. He returned to France and was imprisoned in Rouen charged with 'insoumission'. Following various adventures he managed to leave for the USA in 1940 where he stayed for some years.'[23]
- Charles Philippe Leblond FRS 'escaped from France and moved with his family to McGill University in Montreal, where he became a full-time faculty member of the Department of Anatomy.'[24]
- Guido Pontecorvo FRS 'was arrested as an enemy alien and interned on the Isle of Man when Italy entered the War (in the summer of 1940). After six months he was examined by a magistrate who asked questions, mainly about his attitudes, and released him.'[25]
- Hans Kronberger FRS 'was classified as a 'friendly enemy alien' and dispatched to internment first on the Isle of Man and then in Australia.'[26]
- Heinz London FRS 'for a few months during the summer of 1940 he was interned on the Isle of Man. The development of the atomic bomb was regarded as a permissible occupation for German refugees, London joined the group separating isotopes. He became a British subject in 1942 after which his main work was to investigate the possibility of isotope separation by ionic migration in a liquid electrolyte.'[27]
- James de Graaff-Hunter FRS 'sailed from New York in the SS Zamzam for Cape Town with his wife and eight year old daughter (March 1941). Unfortunately on the 17th April the Zamzam was intercepted in the South Atlantic by the German surface ship Tamesis and the passengers transferred to another ship, the Dresden, and taken to Bordeaux. From there they were transferred to Wesermunde. He was then separated from his family for two months in Stalag X and then for a year at Stalag VIII. In November 1942, largely it is believed through the efforts of Dr N.E. Norlund of Denmark, he was released. A series of journeys through Palestine, Greece, Jerusalem, Baghdad took place followed by an air flight to India. He was appointed Assistant Surveyor General (Technical) until December 1943 and then returned to Dehra Dun to preside over the Department's War Research Institute.'[28]
- Jean Leray ForMemRS 'was a reserve officer. When the War broke out he was called to active duty. Leray subsequently became a prisoner-of-war and was sent to a prisoner's camp in Austria (Oflag XVII A) on 2 July. He remained there for almost five years, until the camp was liberated on 10 May 1945.'[29]

- Knut Schmidt-Nielsen ForMemRS 'was in Norway in the spring of 1940, and managed to get back to Denmark, even though both countries were occupied by the Germans. Schmidt-Nielsen played a role in the Danish underground, helping to manufacture weapons in the machine shop of the laboratory.'[30]
- Lise Meitner ForMemRS 'was a German Jewess who fled Germany to the Netherlands, thence to Denmark (staying with Neils Bohr) before travelling to the Nobel institute for Physics in Stockholm where she spent the rest of her working life.'[31]
- Sir Hermann Bondi FRS 'moved to the internment centre in Huyton, near Liverpool, then to the Isle of Man, back to Liverpool and finally they were transported to Canada, being housed at Camp L on the Plains of Abraham in Quebec (he met Tommy Gold FRS and Max Perutz FRS whilst interred). Bondi was released in June 1941 and he eventually returned to Cambridge in August 1941. He was a mathematician who helped develop radar and served as Chief Scientist to two UK government departments and well as Master of Churchill College, Cambridge.'[32]
- Sir Ludwig Guttmann FRS 'founded of spinal cord injury treatment, the Paralympic Games and the National Spinal Injuries Centre at Stoke Mandeville Hospital.'[33]
- Sir Michael Francis Addison Woodruff FRS 'was a Captain in the Australian Army Medical Corps in Singapore General Hospital shortly after arriving the allied troops surrendered to the Japanese. Woodruff and his medical colleagues along with their patients were marched to the Changi prisoner-of-war camp to join some 55,000 other Australian and British troops as prisoners of War.'[34]
- Sir Rudolf Peierls FRS 'taught theoretical physics at Birmingham and Oxford and was involved in both the development of atomic weaponry and the Pugwash anti-nuclear movement.'[35]
- Valentine Louis Telegdi ForMemRS 'entered Switzerland illegally, and interned in refugee camps for three months, after which he went to live with his father in Lausanne.'[36]

Of course the Fellows and Foreign Members listed above represent just a small fraction of the overall Fellowship who were either interred or displaced. Most who escaped did so with the minimum of publicity. The following might be the exception proving the rule.

The following speech was made by Sir Winston Churchill in closed session.

Note of the secret session 27th June 1940

As we are currently in secret session I am able to tell the House of a very successful piece of work by two officers of the Ministry who were in Paris as Liaison Officers with the Ministry of Armament. These officers, with the cooperation of certain patriotic Frenchman and the British embassy, succeeded in obtaining a ship and arming against attack from the air. This ship was loaded with, among other things, machine tools and large quantities of valuable and secret stores, some of them of almost incalculable scientific importance. There were also embarked on this ship, owing to the efforts that were made, a considerable party of key personnel consisting of eminent scientists and armament experts.

In spite of attempts to bomb it, this ship arrived safely in England and arrangements are being made for the personnel to continue their work in the service of the Allied cause and the stores have been safely disposed of.

Although I cannot do so publically I should like to pay tribute to the highly successful efforts of the representatives of the Ministry and also to the members of the British embassy and to the officers and crew of the ship. A considerable service has been rendered to the Allied cause by the safe arrival of this ship.

Sir Winston Churchill FRS[37]

It is worth including the full report of the Suffolk/Golding Mission as it became known, as an example of the types of activities leading to so-called 'enemy aliens' working in Britain.

Report on Suffolk/Golding Mission

On Saturday, 15 June, after consultation with Major Golding we decided that the situation was sufficiently grave to necessitate our proceeding to Bordeaux. Having done so we reported to the Ministere de l'Armement at 56 rue Commandant Arnould, where we assisted at a meeting with Monsieur Dautry, then the Minister of Armament.

In view of the gravity of the situation Monsieur Dautry decided to save everything he could from the clutches of the Germans and he decided to send two Missions, which had been composed by General Martingnon and General Blanchard to the USA.

In the course of conversation with Monsieur Dautry we ascertained that the Ministry of Supply was in urgent need of technicians. Since telephonic and postal communications were in a state of the most utter chaos there was no possible chance of confirming this with the Ministry, so we decided to proceed on our own initiative. We therefore requested of Monsieur Dautry as many armament experts and scientists as he could spare, which he gladly and willingly accorded us, at the same time giving us as many machine tools as we could take away in the ship. He then instructed Captain Bichelonne, his Chef du Cabinet Technique, to facilitate and expedite these arrangements to the best of his ability, which Captain Bichelonne did.

On the Sunday evening, however, the Reynaud Cabinet, at last fell and the acts of Monsieur Dautry were, in fact, null. In despite all this, however, we proceeded to gather together as many armaments experts and scientists as we possibly could, still aided by Captain Bichelonne, and we commenced the loading of the machine tools.

We should, at this juncture, like to mention that a until representation made by one of us, ie Lord Suffolk, to the highest quarter and in the most uncompromising and bald terms had been made, we met with nothing but a most obstructive and defeatist attitude from the higher members of the Bordeaux Government. However, the demand to Marshal Petain having been made, we succeeded in extracting from him permission to embark the technicians and scientists which we have brought with us.

We then set about acquiring a ship, which said ship was made available to us by the British Embassy, acting through its Commercial Attaché, Mr. Irving, the said ship being the S.S. Broompark, of Denham Ltd, Glasgow, captained by Captain Paulsen.

By Tuesday evening it was obvious that our presence in Bordeaux was known to the 5th Column, since there had been an attempt, which fortunately was unsuccessful, to bomb our ship. However, we proceeded loading machine tools until 5.00am on Wednesday morning, when we decided that for the safety of what we already carried it would soon be as well to weigh anchor and sail for home.

Exclusive of the cargo of machine tools and eminent scientists and armament personnel we had been entrusted by Mr. Irving with carrying a parcel containing over three million pounds worth of diamonds, together with three gentlemen in whose custody this consignment was part. This consignment was sought by the Ministry of Economic Warfare.

Additional to the above cargo we also carried the entire world stock of the commodity known as 'heavy water' (to wit, deuterium oxide, D_2O), together with two members of Professor Curie's Laboratory, who were competent to carry out an extremely important research to which this invaluable commodity was essential.

We also succeeded in extracting from the Naval Authorities in Bordeaux two anti-aircraft 75mm guns, one 'under and over' 9mm pair of Hotchkiss machine guns and one single barrelled Hotchkiss anti-aircraft machine gun of the same calibre. We also succeeded in securing a gun crew of picked members of the French Navy, especially skilled in anti-aircraft defence.

As we previously stated, we weighed anchor at five on Wednesday morning and proceeded to Le Verdon at the mouth of the Gironde, where we took on ammunition for the anti-aircraft equipment, and where we also made one final effort to collect more scientific personnel. This having been done we set sail for Falmouth and arrived there at 8.00 am Friday morning, where we communicated with the Ministry of Supply and having cleared our personal through the formalities of immigration and customs we landed them, together with the diamonds, the various secret papers belonging to various members of the Mission, together with their own, and the 'heavy water' under a guard furnished by the Military Authorities of Falmouth. We then procured a special train, loaded such onto this and proceeded to London, where we arrived at 10.00 am on the Saturday morning. The personnel of armament experts and scientists were then bestowed in the Great Western Hotel and the diamonds delivered to the offices of the Diamond Corporation, whilst the papers and 'heavy water' proceeded, under guard, to a place of security, ie Wormwood Scrubbs.

We should like to make this the occasion for expressing our warmest thanks the following people:

Captain Bichelonne and Colonel Raguet of the Ministere de l'Armement, without whose devoted help, without whose unparalleled efficiency and without whose heartfelt sympathies our Mission could not even have started.

To the members of the British Embassy who were at that time in Bordeaux, with especial reference to the Commercial Attaché, Mr Irving, and the Minister, Mr Harvey, who in the case of Mr Irving, secured us our ship, and in the case of Mr. Harvey secured us an omnibus passport clearance to the customs, without which our task would have been made immeasurably more arduous and difficult.

To Captain Paulsen, the Officers and members of the ship's crew of the S.S. Broompark, who afforded the most loyal, painstaking and hardworking assistance to us and who, in circumstances which might have been of the most extreme discomfort, did all they could to make us and our personnel as comfortable as circumstances would allow.

Thanks are also due to Mr. Barton, who was a Director of the International Chamber of Commerce in Paris, and whom the Embassy asked us to transport as a passenger. Mr. Barton took endless trouble to organise the kitchen department of our journey on the ship and did this very successfully.

Our thanks are also due to Monsieur Berthiez, who then was faced with the removal of dock foreman and dockers organised a scratch crew from the port and with the aid of the First Mate of the ship and Colonel Liebessart, superintended the extremely difficult matters loading these used machine tools himself.

We should also like to comment most favourably upon the efficiency, courtesy and diligence shown to as by Lieutenant Commander Mills RNVR, of the Falmouth Contraband Control. This officer did everything within his power to facilitate landing and despatch of our personnel and valuables and was of the greatest possible assistance to us.

Finally, we should like to cite in the very warmest manner possible the conduct of two secretaries, Miss Morden and Miss Nicolle. Faced with the most uncomfortable possible conditions, faced with hours of work which frequently amounted to some 20 per day, faced at times with the greatest possible danger, they conducted themselves coolly, calmly and extremely efficiently and did what they could to render the operation a success.[38]

This example highlights the need for a process and system of dealing with the incoming scientists and engineers, ensuring that they were placed in the correct environment for their respective skills and knowledge. The National Archives stores a folder containing various instructive letters and minutes concerning this subject.

The first document is from the Committee to Enlist the Help of Refugees Scientists, a Committee sponsored by the Scientific Advisory Committee which was mentioned in figure 1.5, and will be discussed in more detail in Chapter 2.

7th November 1940
War Cabinet
Scientific Advisory Committee
Committee to enlist the help of refugee scientists

A memorandum by Professor A.V. Hill and Sir Alfred Egerton (amended in the light of discussion at the fifth meeting of the Committee held on 6 November, 1940)

1) Terms of reference of the Committee
    It was suggested that a Committee should be appointed with the following terms of reference:
    •   To obtain information as to enemy industries, scientific activities, and other points of interest.
    •   To suggest how certain refugee scientists could best be employed in our war effort.
    This Committee would continue, on an extended scale, the work carried out by Professor Lander's Committee which acted under the aegis of the Air Ministry. Its terms of reference were: …
    to enquire as to the best means of enlisting the help German refugees scientists in the prosecution of scientific work of interest the Air Ministry, and to advise the Air Ministry accordingly …
    This Committee had interviewed 25 refugee scientists and had obtained information of value to the Air Ministry and also to other

departments, before its work was interfered with by the general internment of refugees.

2)   Departmental responsibility
It is suggested that the proposed Committee should be appointed by
the Lord President as a Committee of the Department of Scientific
and Industrial Research, and should report to that Department.
The information gained by the proposed Committee would concern
many different Ministries or Departments, and could conveniently be
distributed to them by the DSIR.

3)   Constitution of the Committee
The Committee should contain a representative of the Advisory
Council of the DSIR and also one from each from the Medical
Research Council and the Agricultural Research Council.
The assistance of the Minister for Home Security would be required in
granting permission to interview internees, internment camps or else-
where, and the Home Office should be represented ...

4)   Registration
The work of the proposed Committee, and of the Aliens Register,
would be facilitated if all aliens professionally and technically qualified
where required to register with the Ministry of Labour and to give full
particulars of their qualifications on the usual card. There should also
be a register for the foreman and artisan class.

5)   Work of the Committee
In the fulfilment of its first terms of reference, it would be the duty of the
Committee to interview those refugees and whose cards indicated that
there might possess useful information. Any information will be reported
to the DSIR, and by them to the Ministries or Departments concerned.
In fulfilment of its second term of reference, it would be the duty of the
Committee to recommend to the Home Office the release from internment of those refugees whom they considered well qualified to give
technical assistance in our war effort, and for whom posts were known
to be available. In the case of particularly well-qualified persons they
would make suggestions as to the ways in which their services might
be utilised, and enquiries as to the possible openings, but they would
not recommend release from internment until that they were informed
that a vacancy existed.[39]

Seven months or so after the above minutes were written, the following
note appeared from the Scientific Advisory Council. The bulk of the note is
not reproduced, however the last page is a list of job titles of refugees who

were unemployed. The required infrastructure needed to collate vacancies with skills must have been large indeed and no wonder that there was a delay in matching vacancies to available people.

4th June 1941
War Cabinet
Scientific Advisory Committee
Enlistment of help of Refugees Scientists
(reference: S.A.C. (40)9th Meeting, Conclusion)
Note by Sir Edward Appleton FRS

At the end of this note there followed a statement accompanying this letter. It reads:

30 April, 1941

Class	Unemployed
Architecture	39
Accountant	1
Actuary	–
Agriculture and forestry lecturers	4
Applied chemistry	107
Artists	6
Barrister	1
Botany	5
Business management	31
Cartography	1
Colder	–
Commodity expert	26
Economics	13
Editor	37
Film production	4
Finance and insurance	1
Fire loss assessors	–
Human pathology	29
Librarians	–
Linguist	532
Mathematics	6
Naval architecture	–
Photographer	1
Physics	25
Physiology	8
Psychology	2

Publicity	–
Pure chemistry	45
Secretary administrator	1
Social administrator	2
Statistician	4
Surveyors etc.	–
Teachers of science	3
Transport	1
University teaching and research	–
Veterinary Pathology	–
Veterinary Surgeons	1
Writer	5
Zoology	5

The numbers of scientific men not been utilised are, with the sole exception of applied chemists, very small. The term 'applied chemists' is, of course, very indefinite and the category may well contain a large proportion of technical rather than scientific men. However, it will be noted that, if any case the records are available to the intelligence branches and I suggest that there is now no strong case before the Committee to recommend the establishment of any additional organisation for interviewing the men concerned.

Clearly action was needed, and started with those refugees with appropriate skills needed for Government projects.

1st August 1941
War Cabinet
Scientific Advisory Committee
Enlistment of help of refugee scientists

Copy of a letter dated 30 July to Lord Hankey from Mr. T.T Scott, International Labour Branch, Ministry of Labour and National Service

I must apologise for being so long in replying to your letter of 2nd July enquiring on behalf of the Scientific Advisory Committee whether this Branch is finding difficulty in obtaining employment for aliens as a result of precautions which seem unnecessarily strict. The fact is that in the meanwhile some aspects of this question have been under consideration between this branch and other Departments and were dealt with only yesterday at an inter-departmental meeting.

There are two principal directions in which precautions are taken in respect of employment of foreigners in this country.

1) Any non-British person desirous of engaging in any form of aliens War service employment must first obtain a permit from the Aliens War Service Department of the Home Office. This requirement is laid down by statutory rules and orders No 1239 of 1940 as amended by No 380 of 1941 made by the Home Secretary. The very wide range of employments for which a permit is required as defined under the Schedules of these Orders

    The granting of a permit is essentially a security matter, depending on the reliability of the applicant and also to some extent on the nature of the prospective employment (eg the secrecy of the work itself or of other work being carried on in the undertaking). The question was recently raised whether some relaxation should not be made in the conditions for granting of these permits, more particularly in the case of friendly aliens of enemy nationality. As a result, at the inter-departmental meeting to which I have referred, certain new arrangements were come to, which, when they are worked out in detail and put into operation should appreciably ease the position.

2) Even in cases where an Allied War Service Permit has already been, or would be granted, the employment of foreigners is sometimes impeded or prevented by apprehensions on the part or potential employers as to the desirability of letting the foreigner acquire the knowledge which would come to him in the ordinary way as a result of and in the course of his employment

    In some cases, these are definitely fears for the preservation of business or trade secrets.

    In other cases, there is no question of such secrets but apprehensions, sometimes, but not always, purporting to be based on security or quasi security considerations, about letting a foreigner acquire knowledge of the general character which might conceivably be misused. However reliable an individual concerned is reputed to be, the policy sometimes followed is to avoid even the shadow of risk by not engaging the foreigner.

    I would add that so far as Allied nationals are concerned, the problem of their absorption into employment is not far from being liquidated and that our main difficulties centre round the Australians and Germans. The rate of progress with these latter groups, though much slower, is nevertheless continuing steadily.

Signed
T.T. Scott[40]

Shortly after this letter was written, Sir Henry Tizard FRS, then Foreign Secretary of the Royal Society, wrote the following:

10th August 1942

Dear Mr. Everett,

I believe that the Scientific Advisory Committee has considered from time to time the question of employment of refugee and alien scientists on Government experimental work.

There is a general opinion in the research sections of the Service and Supply departments that there are alien scientists available of high calibre who should be employed on important experimental work. The research programs at the Experimental Stations are now large, and it is difficult to find suitable experience staff. Would it be possible for the Advisory Committee to look into the problem again?

Perhaps it might help if I quoted two or three cases.

1)   There is a Dr O. Kantorowicz, who is employed at present in the Research Department of the English Electric Company. But so far as I am informed he is not doing work of any great value and his experience leads me to believe that he might well be employed on the development of gas turbines, which is secret work of high importance.

2)   Professor P.P Ewald might also be available. I am told he is at present at Queens University Belfast. He was formerly Professor of Theoretical Physics at the Technische Hochschule, Stuttgart.

3)   Dr Nagelschmidt, now naturalised, is working on soil structure at Rothamsted, and his knowledge and experience would make him suitable for work of much greater importance to the War.

4)   Mr. Fritz Kohl is a German living in this country since March 1935, and is, I am told, a first class scientific man who has had very wide experience of industrial research in Germany. He might be particularly valuable for some War Office work. He is known personally by Professor C.D. Ellis.

There are probably many other examples. In these days, when there is so much to be done and such a shortage of men available and capable of directing younger men, it seems a pity that we should not take advantage of alien scientists provided they are free from suspicion on grounds of security.

Yours sincerely
Tizzard FRS[41]

Mr. Everett was one of the Secretaries of the Scientific Advisory Committee. His reply appears below.

21st of August, 1942

Dear Tizzard,

Sir Edward Appleton has now considered your letter of 10th August, and he inclines to the view that the Scientific Advisory Committee has ascertained both the rule governing the employment of alien scientists and the general principles adopted by the Supply Departments in their employment. We have pressed for the fullest possible employment of such scientists both from the point of view of securing useful information about enemy industry and enemy research projects and for the fullest use to be made of their scientific abilities in our war effort, and we recognise that it is inevitable that they will not always be released except as a last resort and in the case of no equally competent British men had been available.

As I understand it, the question which you raise is not that of finding employment for alien scientists at present unemployed, but seeing that those who are employed are engaged on work which makes full use of their special abilities. This question cannot be treated on general lines and it would seem necessary to follow up in detail the four cases which you mentioned. As a result of our previous enquiries it seems clear that at least in the case of the Ministry of Aircraft Production and the Admiralty, the Supply Departments are prepared to consider the employment of enemy aliens case by case on their merits.

I would suggest that the best way in which we can follow up the cases that you mentioned would be if I could come and see you to discuss the matter.

Yours sincerely
H. Everett
Scientific Advisory Committee
Offices of the War Cabinet[42]

As the process became established, various Government Departments required information on the numbers and occupations of the 'aliens', for example:

28th October 1943
Central register of aliens
Disclosure of statistics

The question arises as to what extent statistics compiled from the Central Register may be disclosed to official and public inquiry. We have before us the following enquiries:

1)   From the Ministry of Information as to the approximate numbers of French civilians at present resident in this country, distinguishing between permanent residents and refugees. Ministry of information had been approached by the National Committee of Rescue from Nazi Terror, for this information (Gen 555/6/3).

2)   From the Ministry of Information as to:
   •   How many aliens there were in this country before the War began.
   •   What is the latest wartime figure.
   •   What are the most 'prolific' nationalities (which is taken to mean which nationalities predominate here), how many there are of each and in what districts they are to be found.
Thus information is required by the compilers of the Encyclopaedia Britannica in America

3)   From the Jewish Refugee Committee as to 'the figures of Germans and Austrians when the yearly census from all over the country was completed in March last' (Gen 301/4/5).
   No definite policy with regard to the disclosure of figures seems to have been laid down and it now seems desirable to formulate some guiding principles. The main point to be borne in mind is that the Central Register records only present registrations with the police, i.e. they are records of aliens who are liable to the provisions of Part II of the Aliens Order and they do not, therefore, include the aliens under 16, those who are in the armed forces and those who are exempt or have been exempted from the provisions of the Order. Furthermore, the records are now of doubtful accuracy, because the police cannot have had the opportunity of checking their registrations during the last few years. There is, therefore, a serious danger that inaccurate deductions may be made from the figures as they stand. In the circumstances, it is necessary, when disclosing any figures, to make it clear that they merely represent the number of aliens registered with the police on a given date.
   It is for consideration whether it is in the national interest to discuss the public detailed information such as the nationalities, sex and

locations of registered aliens. To disclose, for instance, the numbers of Norwegians registered now and before the War might give the enemy some indication of the number of Norwegians who have escaped to this country during the War; and to disclose the number of Norwegian seamen registered in any particular district might be calculated to give the enemy useful information.

At the beginning of the War, the House of Commons was given the figures of registered aliens by nationalities. (Gen 42/1/44)

In April 1941, the Secretary of State gave the total number of aliens registered but said that he did not think that it would be in the national interest to issue a return dividing the figures according to nationalities and sex. (Gen 42/1/58)

In December 1941 the Secretary of State gave the figures have nationals of European states, in bulk and not by nationalities. (Gen 42/1/60)

Whereas there could be no objection to disclosing all pre-war figures, it seems that as a general working rule we should decline, at this stage at any rate, to disclose any up to date figures other than those of a bulk major as has been given in the House of Commons.[43]

2   The Royal Society War Committees

Even within the walls of the Royal Society rooms in Burlington House, little was known at the time of the exploits of either the Royal Society acting as a Corporate Body or of members of the Fellowship acting independently during the War. Yet behind the scenes there was a hive of activity. Not only were certain Fellows working at the highest levels of Government, on some of the most confidential activities of the War, the rooms within the Society were also used to hold confidential meetings.

A relatively short article in *Notes and Records*[1] provided the Fellows with arguably their first glimpse of the scale of the activities. Published after the War this brief article lists the Royal Society Wartime Committees:

- **Advisory Committee on Airborne Research Facilities.**
- **Advisory Committee on Naval Research Facilities.**
- **Committee on Release or Exemption from Internment of Alien Scientists.**
- Committee on the Needs of Research in the Fundamental Sciences.
- **Cultural Relations with India.**
- **Empire scientific conference.**
- **Meteorological Research Committee.**
- **Nature Observations (Army) Committee.**
- Post-war publication of the results of wartime scientific research.
- Prehistoric and archaeological research in East Africa.
- Prehistoric Caverns of Gibraltar.
- **Press and Censorship Bureau.**
- **Scientific Advisory Committee to the War Cabinet.**
- Scientific publications and bookbinding.

There was a lot to accomplish. In many respects the Society suffered from the 'memory of government', or rather the lack of a memory. Through countless initiatives during World War One, individual Fellows demonstrated their resourcefulness and discretion. Yet by the start of World War Two, few Government Officers from the first War were still associated with

Government. There were therefore few individuals able to assess the positive impact the Fellowship and its Fellows could make to the war effort. Arguably Sir Winston Churchill and Lord Hankey were the exceptions, having both served during World War One in various capacities. However, at the start of the World War Two, Sir Winston was not in Government.

Unfortunately, providing a detailed account of each of the Committees mentioned above is not possible, however, it will add value to mention some of them. Those Committees listed above in bold will be mentioned in various Chapters. Figure 2.1 indicates which of the Committees mentioned above in bold were completed during the War, and which were post War activities.

Scientific Advisory Committee

Even before the declaration of War at 11.15am on 3rd September 1939, Sir William Bragg, then President of the Royal Society, wrote the following letter proposing the formation of what was to become the Scientific Advisory Committee.

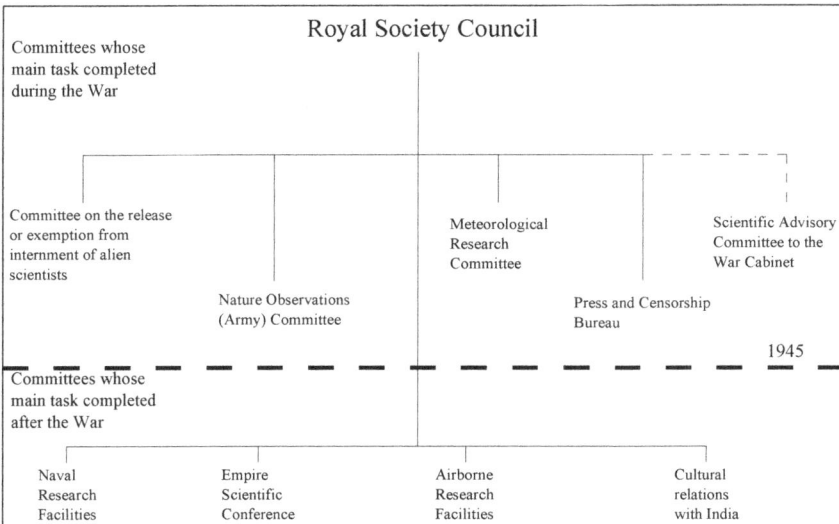

Figure 2.1 Some of the Royal Society War Committees.

13th July 1939
The Royal Society
Burlington House
My Dear Lord Chatfield[2]

During the last 25 years there has been a noticeable improvement in the utilisation of scientific men in the service of the State, and in the application of the results of scientific research by various government departments. I believe, nevertheless, that further improvement is still possible, and I would suggest for this purpose that more use could be made of the body of scientific knowledge available through the Royal Society. I do so now in view of the present urgency in national preparations for defences.

The research organisations belonging to separate departments may easily, to greater or lesser extents, become isolated from one another and from the main stream of independent scientific thought and discovery. Being in each case devoted to a particular service and employing a particular group of men, they may, by reasons of secrecy or departmentalism or merely lack of contact, be unaware of advances in other departments or by scientific people in the world at large. Now it is an outstanding fact that many of the most important additions to and applications of science are due to the combination of knowledge and experience from widely different sources: isolation and departmentalism, therefore, are strenuously to be avoided.

At the present time, it would appear that the most important problems before the Government are those which are directly or indirectly connected with defence; and that the search should be made for all possible means of quickening and improving the application of science to those problems. Further provision of means of contact between the various departments though desirable in itself, would not alone ensure the best result: contact with the main body of scientific workers and of scientific research is also required and is possible only through some independent organisation. I suggest therefore that a very small group of representatives of the various branches of science might be formed from leading members of the Royal Society, who are naturally in close touch with the main body of its Fellows. The duties of these representatives would be:

1) To make themselves aware of what the various scientific organizations under the government were doing; whenever possible to draw attention to knowledge that might be used, or to men who had special ability to deal with any particular matter, and to report thereon to the Committee of Imperial Defence.

2) To advise the Committee of Imperial Defence when desired, as to the best means of obtaining scientific advice or information on any matter,

and to report in general on any deficiencies of which they become aware.

3)   Of their own motion to bring to the notice of the Committee of Imperial Defence any opinion, suggestion or scientific result which might seem to be important to the purposes of that Committee.

In order to fulfil these functions it would be necessary that the proposed Committee should be allowed to learn what the departmental scientific organisations are doing, and to acquaint themselves to some degree with the practical aspects of the problems involved. The Committee would have no executive powers and would not directly originate or supervise research in connection with any Department. Its function would rather be to act as a channel of communication between the government and the scientific community.

Although I am acting on my own responsibility in writing this letter, since the Council of the Royal Society will not meet again till October, I have no doubt whatever from discussions which have taken place at passed meetings that I am saying what the Council would wish me to say.

Details as to the composition of the proposed Committee and other incidental points could be considered later, if and when you thought it would be well to go further with this matter. I would suggest at once, however, that the Committee should be limited to not more than six persons, and that these should not be directly in the employment of any department of the Government.

I know that the Royal Society would be proud to extend in this way its traditional role of cooperation with the Government.

Yours sincerely,
W.H. Bragg
President, RS[3]

Circulated with discretion, yet fairly widely by Lord Chatfield's Department, there was initially mixed response. Figure 1.5 in Chapter 1, showed some of the many Government Committees, the various Directors of Scientific Research of which were often Fellows. Some initially thought involvement by the Royal Society would be counter-productive, adding another layer of bureaucracy. For example:

3rd August 1939
Note on Sir William Bragg's proposal
By Dr H.J. Gough FRS (Director of Scientific Research, Ministry of Supply)

Sir William Bragg's to proposal is stated to have been framed with the following principal object:
To remedy an assumed isolation of service research organisations from:
- Independent scientific thought and discovery.
- Each other.

This is to be effected by a small group of members of the Royal Society, representing various branches of science, who are in touch with the main body of its Fellows. Now, before attempting to provide and apply a cure for any diseased community, it is advisable first to be assured that the community is, in fact, diseased: some enquiry is therefore required into the soundness of the basic assumption of Sir William Bragg.

Starting with the Ministry of Supply technical establishments, it has long been recognised that contact with independent scientific thought and research activity is essential and the method adopted to obtain the corporation of the leading (and more suitable) scientists in this country has been to secure their services as members of, or consultants to, the various technical Committees associated with the work of the technical establishments. The list of Committees is too lengthy to cite in detail (reference may be made, as required, to pamphlet 3/General/378): it will be sufficient for the present purpose to list the scientists at our disposal connection with:
- The Chemical Defence Committee.
- The Ordnance Board.
- R.E. and S. Board.
- Mechanisation Board.

I have prepared such a list and a copy attached as Appendix 1. It will be found to contain 123 names, of which a number no less than 54 are Fellows of the Royal Society. As these particular 54 Fellows are carefully selected because of their association with the particular subjects of interest to our establishments, clearly it is unnecessary to devise any special machinery to put us others in touch with the RS Fellowship.

It may then be argued that, at present, there is no official body of scientists co-ordinating the various research activities of our establishments. This would actually be a true statement but the defect was recognised and pointed out some 11 months ago. For in September last I proposed to D.G.M.P. a scheme for the formation of an advisory council for this very purpose. The scheme received his approval and its implementation has been delayed merely owing to the confusion caused by the change from War

Department to Ministry of Supply. Every member is a Fellow of the Royal Society and all are exceptionally distinguished in those scientific fields of interest us.

I submit that with the formation and functioning of the Council, the Ministry of Supply will have at its disposal the pick of the scientific talent of the country, not only for specialised Committee work and consultation, but also for the most complete co-ordination of these activities.

Then, I feel that the present proposal compels me, very reluctantly, to interpose a personal note. The DSR is himself a Fellow of the Royal Society and, moreover, one who, by reason of his history, has enjoyed most exceptional opportunities of making personal contacts with the leading scientists of both this country and of most European countries and America, all of which he has visited. Since my appointment to the War Office, I have consistently used those contacts to strengthen existing ties with external scientific circles and to create others. It is also my duty to keep up date my knowledge of advances in science and I have endeavoured to do so. As one small result of this, I may point to the various new extra mural researches in current progress. As another example, reference may be made to the very elaborate emergency organisation which has been built up. The extent of this may be gauged by a perusal of Appendix 3 which lists the scientists ear-marked for our use in emergency. It is not for me to dwell on the extent by which this scheme was made possible by my personal acquaintance with, and detailed knowledge of, so many of the individuals named. It will, I think, be clear that such a scheme has only been brought about by personal conversations, correspondence, etc.

Therefore, I would further summit that no external Committee, however distinguished scientifically, is required if its objective is to put our organization in touch with independent scientific thought, as, in my opinion, our required contacts are already firmly established.

The second stated objective of the proposal is to break down the inter-departmental 'wall' which is assumed to exist. Frankly, I know of no such wall. As far as the three DSRs are concerned, we are constantly in touch with each other by formal and informal discussions and correspondence. Our research programs are mutually compared each year, in consultation with DSIR: this is in accordance with a Cabinet instruction. The DSRs serve on each other's Committees and, in fact, every effort is made to avoid any duplication of effort or insufficient pooling of information.

Coming to the highest level of the organization. I attach great importance to the co-ordination of the Aeronautical Research Committee of the Air Ministry and the proposed Advisory Council of the Ministry of Supply. Actually, this must be effected automatically by the common membership of both bodies which will be held by several members. When the time comes, however, I shall propose that, once a year, the two bodies meet for joint consultation on broad scientific lines.

After careful consideration of all the issues involved, I conclude that the creation of yet another body of the type proposed is neither necessary nor desirable. I would add that the basic idea of Sir William Bragg, that the Royal Society should continue its traditional role of co-ordination with the Government in matters scientific is to be welcomed. But his actual proposal probably arises from insufficient knowledge of the really very great extent by which the Services already enjoy that benefit through the valuable individual corporation of the Fellows of the Society. I would suggest that he be informed of the facts.

H.J. Gough FRS
Director of Scientific Research[4]

However, despite this and other objections, the Committee was constituted. Of course the formation will have been helped by the experience of Sir Winston Churchill and his interaction with the Royal Society and its Fellows during World War One.[5] Sir Winston himself was elected Fellow on 29 May 1941 and even prior to his election as Fellow will have built up a first-hand knowledge of the expertise and capabilities of the Fellows of the day.

The following letter to Lord Hankey outlined the formation of the Scientific Advisory Committee, note the delay from Sir William Bragg's initial letter on 13 July 1939 to the formation of the Scientific Research Committee over a year later.

Secret and personal,
20th September 1940
Dear Hankey,

I send you a copy of a minute from Sir Horace Wilson to the Lord President of the Council which he has marked 'I entirely approve of these proposals'.

On Sir Horace's instructions I am in the course of writing to the Prime Minister's Private Secretary, Seal, asking him to put the proposal for the erection of the Scientific Committee before the Prime Minister for approval with a view to early publication and am at the same time sending him particulars of the Lord President's proposal with regard to securing your assistance in the running of the Government research departments for the Prime Minister's information.

There may be an interval of a few days before everything is smooth but I have no doubt in my own mind that we shall get everything we want if we wait for it. I will send you a copy of the letter that ultimately goes to Seal. At present Barlow is sitting on it because he has some doubts about reactions upon the 'Professor'.

I think that the proper step ultimately will be for the Lord President to send for you or write to you as regards your assisting him with the research departments and when he has your formal consent I shall write to the three Secretaries informing them of the new position.

Yours sincerely,
C.A.C.J. Hendricks[6]

Of course the first activity of the scientist is to collect data – it is the life-blood of analysis.

On 14 October 1940, Lord Hankey wrote the following letter to various senior Government Agencies and Departments:

I have been appointed by the Lord President of the Council, with the approval of the Prime Minister, to preside over a new Scientific Advisory Committee. Amongst its terms of reference are:
a) To advise the government on scientific problems referred to it.
b) To advise government departments with regard to the selection of individuals for particular a lines of scientific inquiry or the membership of Committees on which scientists are required.

I fully realise that many of the Government Departments have a highly organised system of scientific research staffed by distinguished scientists. In order to enable me to carry out the terms of reference submitted to the Committee it would be of the greatest assistance if you can furnish me with a short statement of any scientific activities carried on by your department. I am, of course, most anxious not in any way to interfere with the scientific work of the various services, but a short summary would be most helpful to me.

In the case of the Department of Scientific and Industrial Research the Medical Research Council, and the Agricultural Research Council, I have suggested that their synopses might include the following parts:
I.   The terms of reference and membership of the council itself.
II.   A list of Committees and Sub-Committees containing particulars of members and terms of reference.
III.   A list of research establishments maintained a) by the Council, and b) under the supervision of the Council.
IV.   Research activities undertaken outside the Government on behalf of the Council.

I realise, of course, that the suggestions may not be appropriate for the scientific work of your Department, but I mention them in order to show the lines on which I am working.

I am sending a similar letter to a number of other Government Departments.

Signed
Hankey[7]

Following Lord Hankey's appointment, Sir John Anderson wrote to Sir Henry Dale, (see figure 2.2), asking if he would act as Vice-Chairman of the Scientific Advisory Committee. By this time Sir William Bragg had stepped down from his term of office as President of the Royal Society, and Sir Henry elected as President. Of course *all* Presidents led interesting lives. Figure 1.3 in Chapter 1 mentioned Sir Henry's Nobel Prize (1936, Physiology or Medicine), he was also awarded two distinguished medals, namely the Royal Medal 1924 and the Copley Medal 1937. Along with these academic awards he was appointed CBE (1919), Kt (1932), GBE (1943) and OM 1944.

He served on the Member, General Medical Council (1927–1937); the National Institute for Medical Research (1928–1942); was a Trustee of the Wellcome Trust; a Trustee of the British Museum (1940–1963); a Member of the Medical Research Council (1942–1946); President of the British Council (1950–1963); Chairman of the Wellcome Trust (1938–1960) and Director of the Royal Institution.

Figure 2.2
Sir Henry Dale. (© Godfrey
Argent Studio. Reproduced
with permission)

And yet members of the British public would walk passed him not knowing of his occupation or his contribution to the war effort.

The first meeting of the Scientific Advisory Committee took place on 10 October 1940. The minutes of the first meeting appear as Appendix II.

By contrast, an undated, yet lengthy document written by the Parliamentary and Scientific Committee contains the following. Interestingly Professor A.V. Hill has hand-written at the top of this document 'Sir Henry Dale PRS, the P.T.S. first shot!'

> In these circumstances it would appear that the initial effort of the scientific general staff set up in wartime should be:
> a) The co-ordination and exchange of information: the organisation of the distribution of information in regard to our own efforts, and such information as reaches us in regard to those of the enemy and to advise the scientific department as to the scientific possibilities and probabilities of work in any particular direction.
> b) This body would also be invaluable in advising the various establishment departments of existing Ministries as to the qualifications of their future and even their existing scientific personnel and further on the general scientific organisation of those departments. If the government makes proper use of such advice excellent results should flow very rapidly, but of course if the Civil Service machine were to resent or to resist such advice it might become necessary at an early stage to give the general staff actual executive powers with regard to the appointment of personal.
> c) All existing Government Scientific Committees should report to the scientific general staff and scientific general staff should be responsible as stated above for the dissemination of proper information. The general staff would also have a duty of advising the existing scientific Committees as to the direction in which they might work and of recommending to them firms and individuals who might prove to be the most suitable in carrying out the active research work for them.[8]

By 1941 the Scientific Advisory Council (SAC) was in full swing. Figure 2.3 outlines the topics discussed in their minutes.

Date of meeting	Subjects Discussed
10 Oct 1940	Opening remarks etc. These minutes have been reproduced as Appendix II.
15 Oct 1940	Terms of reference, compilation of survey of scientific activities in or supported by the Government, publicity.
21 Oct 1940	Work of the Research and Experiments Departments, intelligence, camouflage, bomb types. Panel for the cooperation with the Defence Services, cooperation with American scientists, employment of scientific and technical refugees.
31 Oct 1940	Government organisations for scientific research in the U.S., meeting of the Defence Services Panel, board of education physical training scheme, application of psychological methods in relation to certain War problems.
6 Nov 1940	Utilization of university scientists in our war effort, possibility of employing scientific and technical refugees.
12 Nov 1940	Committee to enlist the help of alien scientists, suggestions made by Sir Henry Tizard at the third meeting of the Defence Services Panel.
21 Nov 1940	Scientific cooperation with Canada and the US, Sir William Bragg's term of office as a member of the Committee.
24 Feb 1941	Science and post War construction, chemistry and the War, yeast report on an nutritional value and manufacturing
17 April 1941	Research into alternative uses for raw materials and foodstuffs in excess supply, work of the research department of the ministry of supply and importance of training chemical engineers.
26 May 1941	National health and nutrition, research on alternative uses for surplus products, offer of hospitality for British scientists from the Commonwealth Government.
1 July 1941	Alternative uses of surplus products, the enlistment of help of refugees scientists, staff of the central scientific office Washington, visit to this country by Dr Cushing, production transport and storage of food.
8 July 1941	Scientific liaison with Australia, Central Scientific Office
19 Aug 1941	Utilization of the services of research workers in wartime, staff of the Central Scientific Office Washington, photography in the fighting services, and enlisting help of refugees scientists, establishment in the U.S. of an office of scientific research and development, alternative uses for surplus products.
27 Aug 1941	Scientific liaison with Canada, scientific liaison with Russia, the supply of scientific literature from enemy and enemy occupied countries.
10 Oct 1941	Scientific cooperation with Russia.
4 Nov 1941	Science in India.

Date of meeting	Subjects Discussed
14 June 1944	Suggested organisation for combined Services Medical and personal research work, proposals for the development of medical and physiological research and the Royal Navy, the needs of the army in respect of a permanent medical research organisation, proposal for the foundation of an institute of aviation medicine.

Figure 2.3 Some of The Scientific Advisory Committee meetings. (Some examples from National Archives documents CAB 90/1 and CAB 90/2.)

Unconnected with his work as Chairman of the Scientific Advisory Council, on 3 March 1942, Lord Hankey wrote to Sir Winston Churchill offering his resignation as Postmaster General. He also commented on his reluctance to continue his work with the Technical Committees without being in a position with Ministerial rank. In a further letter in the same National Archive folder, Sir John Anderson wrote:

6th March 1942
Dear Hankey,

I am afraid that I have no option but to take your decision to resign from the Chairmanship of the Scientific and Engineering Advisory Committees as final. And acting upon that very regrettable conclusion I have invited R.A. Butler to preside over these two Committees in your place.[9]

The appointment of Mr R.A. Butler as Chairman of the Scientific Advisory Committee continued for some eighteen months, after which the constitution was changed. Sir Henry Dale, commented on 13 September 1943:

after a period with Mr R.A. Butler as Ministerial Chairman in succession to Lord Hankey, it underwent a change of constitution; at the present time the Lord President of the Council acts as President of the Committee; attending and presiding for special purposes, the President of the Royal Society being Chairman at the regular meetings of the Committee.

In a further attempt to keep abreast of scientific events during the War, Sir Henry wrote:

The Royal Society,
Burlington House
London
9 July, 1942
To Mr. Law,

I find that my Secretary has made with yours an appointment for me to see you Tuesday, July 14th at 12:15 pm.

It may be useful for me to tell you shortly the matter on which I should welcome an opportunity of brief discussion. Since I became President of the Royal Society, it has been suggested by my fellow Officers and the Society's officials, that enquiries from the Foreign Office, have been going elsewhere. There may be nothing in the suggestion, but I should like to be sure that there has been no misunderstanding, particularly with regard to the Royal Society's relations to some other bodies which have a more direct responsibility to the Government. My friend and fellow officer of the Royal Society, A.V. Hill, suggested that the matter might be raised by me in an interview with you; I can make clear what is in our minds more easily in about a few minutes of conversation, than in much official correspondence.

I am grateful to you the opportunity.

H. Dale
President, Royal Society[10]

Born on 26 September 1886, and elected Fellow on 2 May 1918, Archibald Vivian Hill FRS, (see figure 2.4), was awarded the Nobel Prize (Physiology or Medicine) in 1922, the Royal Medal in 1926 and the Copley Medal in 1948. These academic honours complemented by OBE (1918), CBE (1946) and CH (1946).

Hill was Royal Society Foulerton Research Professorship (1926–1951) and served on the RS Council on a number of occasions (V-P 1943–1944, 1945–1946; For Sec 1945–1946; Sec 1935–1945).

The following is an extract from a letter forwarded by A.V. Hill FRS and is some collated thoughts of a Fellow intimately connected with War research:

Joint Technical Board
I have received the following letter from a scientific colleague (FRS) who has been working for some years, with great distinction and effect, in a Service Department. The observations contained in it seem to be very pertinent in connection with the present discussion

Like many others I have seen at close hand an alarming number of failures (usually failures of omission) in development work, and have had the

Figure 2.4 V. Hill. (© Godfrey Argent Studio. Reproduced with permission)

opportunity of tracing some of these failures to their sources. There may be varying views how we can best improve our technical development organisation, but everyone seems agreed to that some way of eradicating the most consistent causes of failure is long overdue, and that effort expended to accomplish this will pay a good dividend.

We who have been associated with pure and applied research in recent years have learnt that progress is made by the application of certain 'rational and intellectually honest' principles which are found particularly in scientific circles where work is carried on in the open; to those of us who have scientific training, many of these principles seem just common sense.

AV Hill, 24 July, 1942[11]

This document highlights nine areas of potential improvements to the research and development process.

Once the Scientific Advisory Committee had established itself, there was a steady stream of requests of one sort or another. For example the letter below to The Royal Society, Burlington House, dated 27 August 1942, from A.V. Hill, suggests some names for a vacant appointment.

Dear Everett,

Here are a few names suggested for the Directorship of the Colonial Products Research Institute.

Professor H.J. Channon, Department of Biochemistry, The University, Liverpool

Dr P. Eggleton, The University, Teviot Place, Edinburgh

Dr W.E. van Heyningen, School of Biochemistry, Cambridge

R.A. Kekwick FRS, Lister Institute, London SW1

Professor G.F. Marrian FRS, Department of Medical Chemistry, The University, Edinburgh

T. Moran, British Flour Millers Research Association, St. Albans

Sir William Slater FRS, Dartington Hall, Totnes, Devon

Dr E.C. Bate-Smith, Low Temperature Research Station, Cambridge

Sir Frank Young FRS, National Institute for Medical Research, London

The above were all obtained by looking through the list of the Biochemical Society. Most of them are comparatively junior but it does not look as though we are likely to get anybody of the standing of Albert Chibnall FRS whom we considered last winter.

The field is not one with which I am well acquainted. It might by a good thing to enquire from Sir John Fox FRS, the Government Chemist, who is now I believe Chairman of the Institute of Chemistry. Since there are no chemists on the Scientific Advisory Committee it might be a good thing anyhow to ask Fox to come in to discuss the question with us when we meet.

Yours sincerely
A.V. Hill FRS[12]

Other input covered overseas Institutes, for example:

Draft letter from the Rt. Hon. R.A. Butler MP to the Rt. Hon Oliver Lyttelton MP

10 September, 1942

The Scientific Advisory Council of the War Cabinet, of which I am Chairman, at its meeting yesterday received a report from Professor Egerton, who as you know has served for some months as Director of the British Central Scientific Office in Washington.

As a result of our discussion we think that for the future the Office will work to the best advantage if it is placed under the control not as hitherto of a single Director but of a staff of four, which might suitably be composed of an Administrative Secretary, (who in case of disagreement would have the last word subject to any directions he might receive from the North American Supply Committee) a general physicist, a expert in radar and a chemist.

We might suggest that the post of Administrative Secretary be filled by Dr T.M. Herbert, Director of the Research Dept of the London Midland and Scottish Railway Company, or failing him, by Dr H.J. Ellingham, the Secretary of the Imperial College of Science and Technology. For the general physicist we would suggest Dr P.B. Moon of the Physics Department of the University Birmingham, or failing him, Mr J.A. Hall of the National Physical Laboratory. We consider that the most suitable appointments for the expert in radar and for the chemist would be Mr E.G. Bowen and Mr J.F. Wolfenden.

We consider that in view of the recent appointments to your staff of three Scientific Advisers, it would be appropriate that administrative and technical liaison between the British Central Scientific Office and this country should in future be maintained by your Ministry. I should add that we shall be very ready to continue to assist by suggesting the names of persons suitable for appointment staff of the Office in accordance with our terms of reference.[13]

There are many other examples of the work of this Committee, most activities from which were whilst working at the highest level of Government, some even involved post War infrastructure rebuilding. The following is a letter from Mr R.A. Butler to the President of the Royal Society dated 17 June 1943:

Dear Dale,

I promised to write to you about the question we discussed last week of the supply of Scientific Equipment to enemy-occupied countries.

The Allied Ministers of Education, who meet under my Chairmanship, have two main points of mind. One is that Scientific Equipment &Materials looted by the enemy must be restored; this view will no doubt in due course be put to the proper authorities. The other point is that the Allied Governments do not want to have to depend upon Germany for future supplies as they have had to depend on Germany in the past.

As a preliminary provisional measure, we have asked the Allied Governments to let us have a list by 1st July an approximate estimate of their probable needs. I doubt very much whether these estimates when we have received will be very informative. But there is no doubt at all that they are looking to us and to America to supply the future requirements.

I should be glad if, as you suggested, you would refer the question to the Scientific Advisory Committee.

Yours very sincerely,
R.A. Butler[14]

There followed a report from the British Association Committee on the supply of apparatus and chemicals to universities destroyed by enemy action. The first paragraph indicates the desired outcome.

At a meeting of the Sub-Committee of the British Association considering the restitution of books and apparatus to the universities destroyed or despoiled by enemy action in Europe and China, the Chairman, Professor Timmermans of Belgium reported on the importance of insisting that Germany should be required by the terms of the armistice or of the peace treaty to replace any apparatus and research chemicals which had been taken from occupied countries. If this were said to be impossible, Germany should be required to replace them by other apparatus and chemicals chosen by a scientific Committee of the Allied countries. It would be impossible he thought to carry on instruction in science in secondary schools and universities unless they have the proper equipment and chemicals. The outlay would be comparatively moderate, without essential apparatus, including microscopes, the devastated countries would be severely handicapped.[15]

The work of the SAC even involved scientific work within the Civil Service.

The Royal Society
Burlington House
London
9th July 1943

Dear Dale,
I have looked to the minutes of the Scientific Advisory Committee, particularly the ones relating to Needham's function. It seems to me that no exception could be taken to the second paragraph relating to the transmission of secret military information.

I have spoken to Gough about the possibility of the Scientific Advisory Committee looking into the question which we discussed at the last Council meeting. He was delighted with the idea that I will go ahead and collect the papers together and send them to Allen so that the matter can be put on the agenda of the next S.A.C. meeting. If we then decide to go further we might perhaps invite Gough and Linstead to a later meeting.

I then asked E.W. Salt MP, the chairman of the Parliamentary and Scientific Committee, if he would like the S.A.C. to consider the report on scientists in the Civil Service. He said, as I knew he would, that he would be very pleased indeed. I have asked him, therefore, to let us have a dozen copies, which I will forward to Allen explaining what it is about ...

Signed A.V. Hill[16]

Shortly after the War ended, and roughly the time when Sir Henry Dale stepped down from the Presidency, he sought a meeting with the Lord President of the Council seeking clarity on the future of the Scientific Advisory Council. The following were notes taken at the time:

27th November 1945
Dear Sir Alan,

As you know, Sir Henry Dale came to see the Lord President on Monday to discuss the future of the Scientific Advisory Committee.
I was present at the discussion, and the conclusions reached were:
1) That the Scientific Advisory Council should not be reconstituted for the moment despite the fact that Sir Henry Dale and Professor A.V. Hill are retiring from their Royal Society offices.
2) That the new Committee on Scientific Manpower should be asked to consider the constitutional position and the present composition of the Committee and, if necessary, to make recommendations for its reform.
3) That while this review is taking place, the existing Committee should dispose of any questions which had been referred to it and are still outstanding.[17]

Appendix V documents a more complete assessment of the future of the Scientific Advisory Committee.

Press and Censorship Bureau

In a statement to the House of Commons dated 3 October 1939 the Prime Minister commented, as reported in Hansard:

Finally, I have a statement to make about a matter which has attracted considerable attention and given rise to anxiety in the House and the country – the work of the Ministry of Information. As I stated on Thursday last, the responsibility of the Ministry for news has not extended in any way beyond the provision of means for its communication to the Press. A review has now been made of the arrangements for the distribution of news, including the question of direct contact between the Press and the Departments.

As a result it has been decided that there shall be a reversion to the practice existing prior to the outbreak of War, whereby the Press representatives had direct contact with the various Government Departments, those Departments themselves to furnish the Press and all official communications in such manner as was found most appropriate. Each Department will therefore now make its own arrangements for communicating news to the Press representatives.

It has, however, been represented to the Minister that the mechanical facilities provided in the building occupied by the Ministry other receipt and distribution of government communications have proved a great convenience to a large section of the Press. There is no present intention to do away with these facilities, which will continue to be utilised as a channel by which official communications will be issued to the Press. In any case in which official communications are made direct by Departments, the same communications will be issued simultaneously through the central channel.

The censorship

I should add that some difficulty must be expected in connection with the supply and distribution of news in times of War: the revised arrangements now proposed are intended to reduce difficulties to a minimum, and I feel sure that I can rely upon the press to co-operate in this smooth working. The new arrangements will come into force on Monday morning next.

As regards the censorship, which is at present housed in the Ministry of Information, the responsibility for censorship must rest, like the responsibility for news and its distribution, upon the Departments which are concerned with the subject matter. Departments will, of course, exercise their own control as regards the news that they give out. The central censorship, which is concerned with material or messages submitted by the Press (voluntarily in the case of matter for use in this country) will, as before, by operated by a Censor Officers guided by the directions all the Departments.

In the event of questions arising as to any particular censorship operation, the Minister in charge of the Department affected will answer for it in Parliament.[18]

Sir Walter Monckton was appointed Controller of Censorship. At the same time he supervised the arrangements for the central communication are used to the Press.

5th October, 1939
Dear Sir,
Admiralty and Air Ministry:
Future Press Arrangements

With reference to the Prime Minister's statement in the House on 3rd October, arrangements have been made by the Admiralty and the Air Ministry to give effect to the Government decisions regarding direct contacts with the Press, news distribution by official announcement and censorship. They will come into operation as from Monday, October 9 and are follows.[19]

The National Archives hold several documents relating to press censorship during the War, which at times was somewhat strained between members of the press and Sir Walter Monckton. The aspect of interest here is that involving advice on censorship of scientific journals and papers. At the advice of the Royal Society a panel of advisors was instituted which was a key element in the timely dissemination of relevant information.

Moving to some of the other RS Committees mentioned earlier. Involvement by the Royal Society on the release or exemption from internment of alien scientists was covered in Chapter 1. The Advisory Committee on Airborne and Naval Facilities, the Cultural Relations with India Committee and that for the Empire Scientific Conference will be covered in Chapter 10. The time-scale for the outcome from these Committees was, in the main, post War, even though in some cases the preliminary work started earlier.

Nature Observations (Army) Committee

The basic premise for this Committee was simple. It involved those members of the armed forces willing and able to undertake observations of the flora and fauna surrounding them whilst they were waiting for something to happen. It was a joint Committee between the Royal Society and the War Office. *Notes and Records* commented:

perhaps more than at any time in its history, a British Army during the recent War was simultaneously spread over many quarters of the global. A progressively increasing number of its members were fully engaged in a major task of fighting the enemy. But for many others it was a case of watching and waiting, on anti-aircraft and coast defence duties, in this country and abroad; a form of service equally essential, and liable to boredom. Taking advantage of the two simple facts that there are things of interest everywhere to study if only they be pointed out, and that in the far flung Army there should be some whose interest only requires guidance to be awake, the experiment was tried by bringing some of these phenomena and pairs of eyes together.[20]

It worked and led to many interesting observations.

Other work

Further work took place which involved senior Fellows, without the formation of a formal Committee, for example:

Report of the British
Commonwealth Science Committee
Set up in October 1941 under the
Chairmanship of the President of the
Royal Society

The report covered:
The Bureaux of Information
Collaboration in Research
Organisation of Collaborative Research
Flow of Personnel
Recommendations

The following comments from the introduction set the scene concerning the formation of the Committee and its proposed outcomes:

> The scientific problems provided by the War in connection with technical devices and weapons, supply, medicine, public health, agriculture, food, communications, etc. have brought scientists from all parts of the British Commonwealth into close cooperation. The efficient organisation of this work had necessitated the presence in London of scientific representatives of the various Dominions, that the opportunity was taken by the Officers of the Royal Society of instituting an informal meeting ground for the consideration of joint problems. The general objectives were to secure scientific cooperation in tackling the emergency problems of the immediate post-war period and to ensure that the most should be made of our common scientific resources after the War, for improving both scientific knowledge itself and the life of the peoples of the Empire.[21]

A conference was called at the Royal Society's rooms at Burlington House on 7 October, 1941, in order to discuss the question in general; and, if thought fit, to appoint a Committee, consisting of United Kingdom and Empire representatives, to consider means of promoting cooperation between the several parts of the Empire:

a)   In scientific research.
b)   In the application of science to technical, biological, medical and economic problems.

Representatives of the Dominions and India, representatives of the Royal Society, the Secretaries of the Research Councils, Lord Hankey and others were invited to the conference.

As a result of a small informal Committee was set up under the chairmanship of Sir Henry Dale, the following also agreeing to serve:

Sir David Chadwick	
Dr L.E. Howlett	
Dr H.C. Webster	Australia
Mr. Neville Wright	New Zealand
Lt- Colonel Sir Basil Schonland FRS	South Africa
Dr T. Quayle	Office of the High Commissioner for India
Professor A.V. Hill FRS	Royal Society
Sir Alfred Egerton FRS	Royal Society
Dr Alexander King	Secretary

At a later stage in the deliberations of the Committee a representative of the USA was invited to participate and various offices of the O.S.R.D. have attended meetings. Mr. Eastwood, Secretary of the Colonial Research Committee, and Professor W.E. Le Gros Clark FRS, a member of the Council of the Royal Society, attended the last few meetings.

This report summarises the discussions which have taken place at the twelve meetings of the Committee. The discussions have purposefully been of a somewhat general and exploratory nature. They have, however, been focused mainly on:

1)   Schemes for collecting and disseminating information, for example on the model of the Imperial Agricultural Bureau.
2)   Schemes for collaboration in research within the Empire.
3)   Schemes for facilitating visits and readier movement of research workers, professors and others within the Empire.

The recommendations from this Committee are also kept at the National Archives.

British Commonwealth Science Committee

1)   Set up in October 1941 by the Royal Society. Recommended:
     i)   Discussion of extension of abstracting and information services
           between English speaking countries.
     ii)   Permanent scientific and technical liaison in London of English
           speaking countries.
     iii) A British Commonwealth Scientific Collaboration Committee.
     iv)   Cooperation with countries outside the Commonwealth.
     v)   Value of personal contacts and exchanges.
     vi)   Sabbatical leave.
     All the suggestion of the S.A.C. These recommendations were formal-
     ly conveyed to Dominion Governments which were generally in favour
     of them (South Africa has not yet replied).

2)   Royal Society
     As a result of Sir Henry Tizard's visit to Australia and to Canada and
     opinions expressed by Dr Marsden and Brigadier Schonland, a scien-
     tific conference of representatives of the Dominions to discuss plans
     for the maintenance of Imperial collaboration in scientific research was
     proposed. The SAC, with the approval of the Lord President recom-
     mended that such a conference be held after the War and invited the
     Royal Society to take steps toward convening such a conference. This
     is now being done. Sir Stafford Cripps has told the Lord President
     that he fears that if such a conference is called by the Royal Society,
     insufficient attention will be paid to the industrial aspects of scientific
     collaboration.

        And yet on 23rd January 1945, Professor A.V. Hill FRS wrote the
     following to Sir Edward Appleton FRS:

        Dear Appleton,
        Thank you for your letter of 18th January returning a copy of the
        report by the Sub-Committee of the Parliamentary and Scientific
        Committee.
           I asked Salt, the Chairman of the P. and S. Committee the
        other day whether he would arrange to keep you regularly supplied
        with their papers as they issued them. He said he gladly would, so
        I hope they will turn up.
           I suppose you have heard that the Treasury has agreed to
        put in the estimates the sum of £15,000 for the expenses of the
        Empire Scientific Conference to be held either next summer, or
        more probably in 1946. Egerton and I have now written to the
        High Commissioners for Canada, Australia, South Africa, New

Zealand and India, letters of which the enclosed copy of the one addressed to Massey is typical. I will write also an appropriate letter to the Colonial Office.

I hope we shall soon be able to arrange the first meeting of the 'Policy Committee' to get on with general plans for the Conference.

Yours ever,
A.V. Hill FRS[22]

3   Government War Committees

There were so many Government Wartime Committees that to list them all, and their membership, would add little value on its own. Figure 1.5 in Chapter 1 details the relationship between various Government Committees and their respective Ministries. Whilst this modified version was sent by Mr H. Everett to Lucker as described in Chapter 1 and Appendix I, the original diagram first appeared in section 8 of the Scientific Advisory Committee minutes.[1] Annex VII of that document is titled 'Explanatory Statement on the Diagram illustrating the Organisation of Research for Defence Purposes in Great Britain', and appears here as Appendix III.

Chapters 6–9 will discuss some of the work undertaken by those Committees relevant to the subjects under discussion. It might be worthwhile here, to highlight some of the more unusual activities of a small number of Committees, for example:

Joint Conference of representatives of the Scientific Advisory Committee (Defence Services Panel) and of the Engineering Advisory Committee

This joint series of meetings were attended by various Fellows. Some of which were:

Sir Edward Appleton FRS, Sir Alfred Egerton FRS, Sir Henry Dale FRS, Lord Hankey FRS,
Professor A.V. Hill FRS, Sir Edward Mellanby FRS, the Rt. Hon Viscount Falmouth, Mr W.T. Halcrow, Sir Clifford Paterson FRS, Sir Harry Ricardo FRS and Sir Henry Tizard FRS.

The minutes of the meeting held in the Privy Council Office on Wednesday 7 January 1942 include the following items:

Methods of detection of a German channel tunnel

The Chairman welcome Dr Bullard and Dr Rankine and invited them to express their opinions on the methods of detection suggested in the papers and minutes, and to suggest any other methods which they consider valuable.

Dr Bullard explained that he had considered the various methods suggested and made a statement in reply, the effect of which was as follows:

a) He took the view that the following methods of detection did not offer any practical solution:

   a) **The Gravemeter.** The probable range of this machine was only some 30 feet through dry land.

   b) **The torsion balance.** This instrument, working on land, might be able to detect a tunnel 16 feet in diameter through 100 feet of chalk, but would probably be incapable of detecting a tunnel at a greater depth than 100 feet. Could not be worked in a submarine or a boat.

   c) **Echo sounding** chalk was an impossibly bad medium through which to use such apparatus.

   d) **The ASDIC** could not be used through chalk.

   e) **Radiolocation** did not appear to present a practical method of detection.

b) It might be possible to detect electric currents in the tunnel

   Direct Current would probably be used. The possibility of detecting this depended on a number of unknown factors such as the current employed and the degree of screening by the tunnel lining. It seemed likely that the fields produced would be less than that discussed in c(i) below. If alternating current was used it might be more difficult to detect, since the instruments would record other electric disturbances of this character. The electricity in the workings might be detected by means of a number of loops of cable laid over the probable line of the tunnel. He was not too optimistic as to the utility of this method of detection, particularly in view of the electric railways near Dover

c) He considered that there were two promising methods of detection, magnetic and acoustic.

   i) **Magnetic.** Vertical magnetometers, such as were manufactured by E.R. Watts and could pack into a suitcase, could be employed to detect reinforcement to a tunnel up to a range of 300 feet or so. Reinforcement round the tunnel would probably be more effective than longitudinal reinforcement. It is considered that it should be possible to detect a disturbance of 10 gamma in the field. When over the tunnel, the magnetic field should increase. If the tunnel was unlined he was doubtful whether the magnetometers would be able to detect any steel rails, pipes etc. He suggested that if the menace was taken seriously a series of recordings should be taken in the near future at intervals of about 30 feet along a line crossing the probable line of a possible tunnel. The work could probably be done in about a fortnight. These readings should be repeated at intervals in order to establish the normal range of variation at a time when it could safely be assumed that no tunnel had been dug. If these readings showed a narrow range of variations, and later on at a particular point recorded an abnormal change a German channel tunnel would probably have been detected. Two instruments would be required, one to make measurements at a fixed point and one to move from point to point along the line.

ii) **Acoustic.** Low frequency listening machines placed in bore holes in chalk near the beach should be able to detect the noises inevitable in drilling a tunnel at a depth of some 200–300 feet and at a distance up to approximately 1/4 mile. While it would be theoretically possible to place listening machines on the bed of the sea sufficiently far out from the shore to be away from wave and shingle noises, the practical difficulties of laying the necessary heavy armoured cable in the face of the German guns would, in his opinion, be insuperable. The Admiralty had staff and equipment capable of carrying out such listening experiments, but it was certain that they would be reluctant to take skilled men off other urgent work. To be effected the acoustic posts should be spaced along some eight miles of shore, preferably sunk into the ground.

d) He assumed that the Germans would follow the lines suggested by the Channel Tunnel Company, or a little to the east of it, both because this was the easiest route and because there exists a considerable body of published data as to the conditions of the seabed along this line. He considered that a deep level tunnel in the Palaeozoic rocks was fantastic. Holes 4 to 6 inches in diameter could be bored through 400 feet of chalk in two or three days by the use of the modern machinery in this country owned by the Anglo-Iranian Oil Company Limited.

Dr Rankine expressed general agreement with Dr Bullard's views, basing his opinion on his experience with the Anglo-Iranian Oil Company Limited, and pointed out that if it could be assumed that no German channel tunnel existed at present, the circumstances were easier than those normally met with in geophysical investigations. If tests could be made before the existence of the tunnel then subsequent detection should be considerably easier than the detection of something underground which existed at the start the investigation.

In the course of a general discussion reference was made to the old Pilot Heading, and it was suggested that experiments could usefully be made over this both by magnetometers and by listing apparatus. If the listing apparatus was unable to detect the explosion of detonators, it would be clear that they would be useless for the purpose in view.

Professor Egerton reported that he had made enquiries into the question of the probable visibility of spoil bumped up to the bed of the channel in the form of slurry. It seemed certain that such pumping in 50 feet of water, or less, would be visible on the surface, and that though it was probable that this would also be the case in deeper water it seemed possible that towards the middle of the channel it could not be detected.

The meeting:

a)   Invited the Chairman to request the War Office Authorities to ask the Chief Engineer at Dover to investigate and report on the present conditions of the Pilot Heading with a view to its use for experiments in detection.

b)   Agreed that a further meeting should be held when this report is received and to decide on trial experiments, and that should the use of the Pilot Heading prove impractical the possibility of using other workings should be considered.

c)   Agreed that the Photographic Authorities of the RAF should be asked to take a series of photographs of the probable site of the tunnel heading near Cape Gris Nez and over the channel along its most likely course.

d)   Expressed agreement with Dr Bullard and Dr Rankine that the magnetic and acoustic methods of detection were the most promising.[2]

The Engineering Advisory Committee in part, owed its existence to the formation of the Scientific Advisory Council mentioned in Chapter 2 and in the above minutes. The process of its formation started in December 1940 with a letter to Sir Winston Churchill who was by then Prime Minister.

23rd December 1940
To the Right Hon. Sir Winston Churchill
Prime Minister,
10 Downing Street
Sir,

We, the undersigned Presidents of Engineering Institutes, wish to put before you a suggestion for increasing the war effort of the nation.

We write as a result of the unanimous recommendation of the Engineering Joint Council and with the full support of our respective Councils. Our institutions have a combined corporate membership of over 46,000 fully qualified professional engineers, including the greater number of these engaged in aeronautical, civil, electrical, marine, mechanical and structural engineering.

This is a War of scientists and engineers but there is frequently lack of understanding as to their respective spheres of activity. In wartime, the scientist is responsible for original research directed to the discovery of a) new means of offence and defence and b) improvements in medicine, agriculture, health and nutrition; the engineer is responsible for applying the discoveries of the scientists in group a) to practical use in warfare by designing and producing plant and apparatus which is robust, reliable, capable of rapid and economic manufacture in bulk and convenient for use by the Fighting Services. A typical example of what can be achieved by their joint effort the development of ASDIC in the last War. In the present War such co-ordination is of the utmost importance.

Characteristics which make the best scientist or the best engineer are seldom combined in one person; hence the successful transition from scientific research and invention to full scale production in the factory is seldom made direct and takes a long time. It is for this reason that industry in peacetime set up large organisations for engineering research and for engineering design and development to bridge and shorten the difficult gap between the scientists and the engineer. The fullest utilisation of these organisations in wartime should be made. Under modern conditions greater and swifter progress can be made by organised effort than would be possible by the efforts of individuals. This is even more so in engineering research and development than in scientific research. The Government has recognised this as regards scientific research and the work of scientists is being coordinated through the strong Scientific Advisory Committee which has recently been appointed. No such action has been taken in the field of engineering.

It is felt by engineers that their industry is not giving such full assistance in the war effort as would be possible if its resources were better coordinated both within the industry and with scientists and government departments. As typical examples of this we know that:

a)   The time taken to put new scientific ideas into practical production and use would be shortened by closer cooperation between the engineer and scientist.

b)   There is a need for closer cooperation between the engineering research and design departments of industry and government scientific establishments.

c)   There are untapped reserves of capacity in industrial and other research laboratories which could be used to supplement the work of scientists and Government Departments.

d) The emphasis on secrecy has tended to prevent engineers from cooperating fully with each other on wartime problems.

e) Industry could assist in obtaining and rapidly training the personal required by the Services for operating and maintaining new types of plant and apparatus.

In order to facilitate a greater contribution by engineers to these and similar problems, our Institutions strongly urge that the Government should appoint, in parallel with the Scientific Advisory Committee, an **Engineering Advisory Committee** with some such terms of reference as the following:

1) To advise the government of engineering problems.

2) To advise Government Departments on the selection of organisations or of individuals for particular lines of engineering development.

3) To bring to the notice of the government promising new engineering developments which may be of importance to the war effort.

4) To advise the Government on methods of initiating and speeding up the solution of new engineering problems arising from the War and on arrangements for rapid production of plant and apparatus.

5) To advise the government of methods of obtaining and training personnel for technical duties in the Services.

6) To appoint, with the approval of the Government, Sub-Committees, to deal with specific branches of engineering and panels to study specific problems.

7) To act in cooperation with the Scientific Advisory Committee.

We trust that you will approve of this proposal. Any reply which you wish to make should be addressed to the President of the Institute of Electrical Engineers, who is in communication with us all.

We have the honour to be
Sir
Your obedient servants[3]

This letter was signed by the Presidents of the following institutions:

The Institution of Civil Engineers
The Institution of Mechanical Engineers
The Institution of Naval Architects
The Institution of Electrical Engineers

The Institution of Municipal and County Engineers
The Institute of Marine Engineers
The Acting President of the Royal Aeronautical Society
The Institution of Structural Engineers

As one might expect various senior personnel from Government Departments expressed opinions about the formation of this Engineering Advisory Committee. Perhaps of interest here are those comments made by Sir Henry Tizard FRS to an individual called Barlow.

3rd February 1941
Dear Barlow,

I return to the document about the proposed Engineering Advisory Council. The engineers have a good case on paper. I entirely agree with their paragraphs a) and b) on page 2. There is no lack of science or new scientific ideas in this War. Our shortcomings have mainly been due to the fact that the engineering of the ideas has not been good enough. It seems to be a much easier to get a first class young scientists than first class young engineers; but that may well be because people like myself know where to look for the first class scientists and are not so closely in touch with the sources of first class young engineers, who do not by any means come only from the universities.

I expect their paragraph c) on page 2 is also correct, but I doubt if the reserve of capacity in the industrial research laboratories this very large.

As for d) it sometimes seems to me that there is no secrecy these days – I hear all our secret devices being talked about quite freely.

As for e) I think we could get, and certainly have got, industrial assistance in training personnel.

Broadly speaking there is just as much a case for an Engineering Advisory Council as for a Scientific Advisory Council. The problem is to choose the right individuals and to settle what they can do usefully without overlapping to too great an extent with other people's responsibilities. I think a suitable Committee could be of much help if they were used as suggested in paragraphs 2, 3, 5 and 7 of the tentative terms of reference on page 2. I am not sure that they could be of much active assistance on page 4, but I really don't know enough about the production problems to express a good opinion.

My advice would be to accept the suggestion to form the Committee and be very careful about the selection of the personnel of the Committee, and, finally, not to worry too much about the terms of reference. If you get the right people on the Committee the exact terms of reference don't matter very much. If you get the wrong people, they don't matter at all because nothing useful will be done.

Yours ever,
Sir Henry Tizard FRS[4]

Born in Gillingham, Kent in 1885, Sir Henry Tizard FRS was Foreign Secretary of the Royal Society and Rector of Imperial College of Science and Technology at the time he wrote the above letter. He was a Lieutenant-Colonel and Controller of Experiments and Research in the Royal Air Force during the latter stages of World War One (1918–19), and so was somewhat familiar with the need for research and development infrastructure. Incidentally, he was also Chairman of the Aeronautical Research Committee 1933–43, a Member of Council of the Ministry of Aircraft Production (1941–43) and a Development Commissioner.[5]

Not that he knew it at the time, but Sir Henry was to be one of the founding members of the Engineering Advisory Council.[6]

Figure 3.1 Sir Henry Tizard. (© Godfrey Argent Studio. Reproduced with permission)

There then followed a flurry of activity designed to determine suitable candidates for the Committee.

Confidential
28th April 1941

Dear Professor Robertson FRS,

I am writing to confirm my telephone message of this morning.

The Lord President of the Council is appointing an Engineering Advisory Committee, with the terms of reference which I enclose, and he would be greatly obliged if you would consent to serve upon it. The chairman is Lord Hankey and the other members besides yourself are Lord Falmouth, Sir Henry Tizard FRS and the following, who had been chosen after consultation with the Presidents of the Institutions of Civil, Mechanical and Electrical Engineers:

> Mr. W.T. Halcrow
> Mr B.W. Holman
> Sir Harry Ricardo FRS
> Dr A.P.M. Fleming
> Mr J.R. Beard
> Sir Clifford Paterson FRS

I mentioned to Sir John Anderson the point which you raised in our conversation, namely, that you did not feel that you possessed yourself an up to date knowledge of engineering works, management or production, and that you felt that someone with this experience was needed on the Committee. He asks me to say that it will be open to the Committee, if they think it desirable, to add to their number someone with the kind of qualifications which you suggest, and that for this purpose it would be open to you or to any other member of the Committee or to the President of the Institution of Mechanical Engineers to suggest names for the purpose; but that both he and Lord Hankey very much hope that you will yourself be able to join the Committee.

They would like to be able to announce the appointment of the Committee in the press on Wednesday morning. If you could let me have a reply tomorrow (Tuesday) morning, I should be grateful.

Yours truly,
J.A. Barlow[7]

The Engineering Advisory Committee was duly constituted and contributed fully within their chosen remit, as well as contributing to the joint

Committee mentioned above in connection with the potential channel tunnel. A brief account of a further two Government Committees is included below showing further facets of Government research and development. The two Committees chosen are the Advisory Council on Scientific Research and Technical Development and the Meteorological Research Committee.

Advisory Council on Scientific Research and Technical Development

This Committee was part of the war effort within the Ministry of Supply. The terms of reference for this Committee were published on 4 January 1940.[8]

1) To consider and initiate new proposals for research and development, and review research and development in progress by the Ministry of Supply establishments in relation to the most recent advance in scientific knowledge.
2) To advise on scientific and technical problems referred to them.
3) To make recommendations regarding the most effective use of scientific personnel for research and development.
4) To report to the Minister of Supply.

The membership of this Committee comprised of some familiar names:

Lord Cadman of Silverdale	Chairman
Professor E.N. da C. Andrade FRS	
Sir Edward Appleton FRS	
Sir Joseph Barcroft FRS	
Sir William Bragg FRS	
Major-General E.M.C. Clarke	Director of Artillery, Ministry of Supply
Sir John Cockroft FRS	
Major-General A.E. Davidson	Controller of Mechanisation Development, Ministry of Supply
H.J. Gough FRS	Director of Scientific Research, Ministry of Supply
Sir Henry Guy FRS	Chief Engineer, Mechanical Engineering Department, Metropolitan-Vickers Electrical Company Limited
Brigadier General Sir H.B. Hartley FRS	Vice President and Director of Research, LMS Railway
Sir Ian Heilbron FRS	Sir Samuel Hall Professor of Chemistry, University of London

Professor A.V. Hill FRS	Secretary, Royal Society and Foulerton Research Professor
Professor R.S. Hutton	Goldsmiths' Professor of Metallurgy, Cambridge
Sir David Pye FRS	Director of Scientific Research, Air Ministry
Sir Robert Robertson FRS	Director of the Salters' Institute of Industrial Chemistry
Sir Robert Robinson FRS	Waynflete Professor of Chemistry, Oxford
J. Rogers	Deputy Director General of Explosives, Ministry of Supply
Sir Frank Smith FRS	Director of Instrument Production, Ministry of Supply; Director of Research, Anglo-Iranian Oil Company
Sir Richard Southwell FRS	Professor of Engineering Science, Oxford
Sir Reginald Stradling	Chief Technical Advisor, Ministry of Home Security
Sir Geoffrey Taylor FRS	Yarrow Research Professor of the Royal Society
Lieutenant General Sir Maurice Taylor	Senior Military Advisor, Ministry of Supply
Sir Henry Tizard FRS	Rector, Imperial College of Science; Chairman, Aeronautical Research Committee
C.S. Wright	Director of Scientific Research, Admiralty

Joint Secretaries

E.T. Paris	Deputy Director of Scientific Research, Ministry of Supply
F. Roffey	Deputy Director of Scientific Research, Ministry of Supply

This Committee had various Sub-Committees including:

- Anti-Concrete.
- Ballistics.
- Communications.
- Explosives Research (Physics and Physical Chemistry).
- Fighting Vehicle Armament Research and Development.
- Flax.
- General Physics.
- Gun Design.
- Propellants.
- RDF Applications.
- Static Detonation.
- Substitutes.
- Unexploded Bomb.

There was so much work from these and other Committees that it is difficult here to take any one topic from one of the Committees and do justice to the technology. Perhaps the following letter from Sir Andrew Duncan to the Committee Chair might help to illustrate the scope and depth of the work undertaken:

Appendix 1 from the minutes of the meeting
from Thursday 25th February 1943
held at Burlington House

My Dear Sir Frank,

I have read very great interest and pleasure the report of the Advisory Council on Scientific Research Technical Development for 1942, covering its activities during the third year of its existence. The review embraces a wide and rich field of scientific and technical effort and clearly reveals great achievement and progress in many directions with most beneficial results to our War efficiency and effort.

Without going into too much technical detail, I would mention certain broad aspects of the report which gave me much satisfaction.

1) The attention given, by scientific study, to the improvement of existing weapons, equipment, etc. As exemplified by the work on such items as large bombs, flash suppression, explosives, bridging, springs, etc.

2) The assistance given to production processes and economy; the studies of fine grain fillings, TNT and RDF, resetting of flax, use of low alloy steels for guns, are typical.

3) New applications of science and technology in a wide field, including: fundamental study of gun design, hollow charge projectiles, static detonation and blast, fuses and fuse setting, RDF and radio communications, devices for dealing with unexploded bombs.

I have also noted with very great interest and appreciation the extremely close scientific and technical liaison existing between the Council and the Dominions and the USA, and the occasions on which the Council has prepared special appreciations for the General Staff.

With my thanks to yourself and the Council for its excellent report and my appreciation of its past work, I send my best wishes for the success of its future activities.

Yours sincerely
Andrew Duncan[9]

Meteorological Research Committee

As with the Scientific Advisory Committee, the origins of the Meteorological Committee pre-date the War. Some of the key milestones in the formation of this Committee were written up on 30 August 1947.

References to some old papers, in file 854507/38, bearing on the early history of The Meteorological Research Committee

29th Nov 1938	at a meeting of the Meteorological Committee Professor Chapman FRS proposed the formation of the Met. Research Committee and the DMO agreed to draft the regulations (E.1A and M.5)
13th Feb 1939	S. of S. Approved the proposal in principle (M.8–10)
9th Sept 1939	Treasury unwilling to agree payment of fees, and in view of the national emergency DMO agreed to defer action to establish the Committee (M.29)
2nd July 1941	Professor S. Chapman FRS re-opened the question of setting up the Committee and submitted a detailed statement of his views and proposals, including the suggestion of fees should be dropped for the time being (E.30A, B)
8th Sept 1941	DMO requested from the DUS authority to establish the Met. Research Committee and submitted a revised draft of the proposed Regulations (M.41, E.41A)
29th Sept 1941	S of S approved the setting up of the Met. Research Committee (M.46)
7th Nov 1941	S of S (Sir Archibald Sinclair) appointed the membership as follows: Professor S. Chapman FRS (Chairman), Sir David Brunt FRS, Dr G.M.B. Dobson FRS, Sir Geoffrey Taylor FRS, DMO, (N.K. Johnson), Representatives of the Admiralty (Captain L.G. Garbett), Air Staff (Group Captain H.J. Saker), and of the D.G. Civil Aviation (C.B. Collins)
10th Dec 1941	the Committee held its first meeting (at the Royal Aero. Soc rooms). Miss Flora Jones was Secretary
30th Mar 1942	a representative of the Ministry of Aircraft Production, viz Sir David Pye FRS, Director of Scientific Research, was appointed to the Met. Research Committee[10]

In a letter dated 15 May 1942, N.K. Johnson, the Director of the Meteorological Office, Air Ministry, commented:

Meteorological Office, Air Ministry
Meteorological Office Orders, Supplement
Meteorological Research

At a meeting of the Meteorological Office staff held in March 1939, I announced that I hoped to be able to develop a measure of organised meteorological research, and that an advisory body, to be known as the Meteorological Research Committee, was being appointed towards that end. The outbreak of War prevented the scheme from being put into operation for a time but about six months ago the project was revived.

The terms of reference of the Meteorological Research Committee are broadly to advise the Secretary of State for Air as to the general lines along which meteorological research should be developed and to assist in the carrying out of investigations and research within the Meteorological Office. It is also responsible for coordinating investigations undertake within the Meteorological Office with related research carried out elsewhere, both in this country and abroad.

The composition of the Meteorological Research Committee is as follows: Professor S. Chapman FRS (Chairman), Dr G.M.B. Dobson FRS, Sir Geoffrey Taylor FRS, DMO, (N.K. Johnson), Director of the Naval Meteorological Service, Ministry of Aircraft Production. Representative of the Air Staff, representative of the Director General of Civil Aviation

The Committee is now holding regular meetings. A research program has been drawn up and work upon a number of problems is in progress. A lot of these problems are naturally directly concerned with the war effort and in consequence have to be treated as secret. Some members of the Meteorological Office staff have already been brought into certain of the investigations, and others will be brought in as necessary. Reports dealing with problems which are of importance or interest to the staff will be circulated for their information.

As a matter of general policy the Committee have decided to concentrate their attention for the time being upon practical and applied problems, and have invited the cooperation of the Royal Society for dealing with investigations of a more purely scientific character. The Royal Society have expressed their willingness to assist in this way, and arrangements have been made for a comprehensive investigation into the various aspects of the problem of radiative equilibrium in the atmosphere. More detailed information about this particular investigation will appear shortly in Nature and in the journals of certain scientific societies.

N.K. Johnson
Director
Meteorological Office[11]

The strategic significance of this and other meteorological activities cannot be overstated. For example, The Met Office, the UK's National Weather Service, post the following on their website:

> The weather was crucial to the Allied Forces's success for the D-Day landings in June 1944. General Eisenhower's chief meteorologist, Group Captain James Stagg, a Met Office forecaster, advised of a narrow 'weather window' for the operation to go ahead: 'probably the only day during the month of June on which the operations could have been launched' President Truman later said.[12]

Chapter 2 listed Royal Society Committees active in World War Two, one of which was the Meteorological Research Committee. For completeness, Appendix VI provides further information of the Royal Society's activities in this area.

The work of the Royal Society outlined in Appendix VI was in response to the Chairman's remarks in the first meeting of the Meteorological Research Committee held on Wednesday 10 December 1941.

> The last 20 years have seen growing recognition of the importance of the adsorption of solar radiation by ozone, and of the close connection between weather conditions and the ozone distribution, despite the extremely low concentration of atmospheric ozone. Ozone, with water vapour and carbon dioxide, is one of the important polyatomic constituents whose absorptive and radiative powers far exceed those of the more abundant atmospheric gases, nitrogen, oxygen and argon. This property of polyatomic gases makes it desirable a) to ascertain what is the detailed chemical composition of air, including even minute amounts of the polyatomic constituents, such as the oxides of nitrogen and compounds of carbon other than CO_2; b) to determine their spectra and absorption coefficients, including those in the infrared and ultraviolet regions of the spectrum; c) to develop practical methods of determining the amounts of the chief polyatomic constituents of the air, both near the ground and at higher levels; d) to improve the theory of the

radiation balance and the atmosphere, with the aid of the data which the investigations a), b) and c) would afford.

Such investigations are essential to a proper theoretical foundation for metrology, and it seems likely that some progress in them is possible even in wartime without calling upon those whose time is fully employed on work of the immediate urgency. Such progress would be greater if co-operation from the USA and the Dominions would be obtained.[13]

By 1943 the Meteorological Research Committee were involved in the following:

I)   Development of Meteorological Instruments
   i)   Development of hygrometers.
   ii)   Improved design of radio-sonde.
   iii) Development of a distance-reading anemometer.
   iv)   Development of a device for giving warning of deteriorated visibility.
   v)   Design of an automated radio reporting station.
   vi)   Design of height of cloud base in daytime.
   vii) Determination of thickness of cloud layer.
   viii) Improvement in measurement of visibility at night.

II) Investigations for Improving Forecasting
   i)   Improvement in forecasting technique.
   ii)   Determination of relationship between stratospheric phenomena and synoptic situation.
   iii) Improvement technique for forecasting particular conditions.
   iv)   Estimation of local and topographical characteristics.
   v)   Development of a method of long range forecasting.

III) Dynamical and Physical Measurements
   i)   Application of upper air data.
   ii)   Prevention of ice accretion.
   iii) Physical characteristics of fog.[14]

The involvement of other Fellows

Undoubtedly, the Fellow with the highest profile during World War Two was Sir Winston Churchill FRS. Born in 1874 and educated at Harrow School, Sir Winston was a veteran of the Boer War and played a significant role in Government during World War One. A year or so after his election as Prime Minister on 10 May 1940, Sir Winston was elected Fellow of the Royal Society.

Of course Sir Winston went on to receive many awards following the War including the Order of Merit in 1946. He was admitted a Knight of the Garter in 1953 and was awarded the Nobel Prize for Literature also in 1953.

The prominent placement of his statue in Parliament Square overlooking the Houses of Parliament is fitting, (see figure 3.2), perhaps even more so as it is but a stones' throw from Downing Street and the War Cabinet rooms, see figure 3.3.

Incidentally, Jan Christian Smuts FRS also has a statue dedicated to him in Parliament Square. One might speculate, however, that the statue erected in his honour was for work accomplished earlier in the 20th Century.

Figure 3.2 Sir Winston Churchill. (Author's photograph)

Figure 3.3 The Churchill War rooms. (Author's photograph)

Some other Fellows not yet mentioned for their contributions to Government Committees or Departments during the War include:

- David Forbes Martyn FRS 'became the first head of the Radiophysics Laboratory in Sydney where he remained until 1941 when he left to establish an Operational Research Group within the Australian Armed Forces. Towards the end of the War he returned to his basic researches on the upper atmosphere at the Commonwealth Observatory, Mount Stromlo in Canberra.'[15]
- David Meredith Seares Watson FRS 'found himself in America as a Member and Acting Secretary of the Agricultural Research Council concerned with the problems of poultry (at the beginning of the War). Early in 1940 he became Secretary of the Scientific Sub-Committee of the Food Policy Committee of the War Cabinet.'[16]
- Ernest Frederick Relf FRS 'was Superintendent of the Aerodynamics Department of the NPL for twenty years from 1925–45. Served on various Sub-Committees of the Advisory Committee for Aeronautics including: Aerodynamics (1918–1966), Stability and Control (1923–1945) and the Fluid Motion Panel (1931–45). He also served, at various times and for various relatively shorter periods on the following bodies: Wind Tunnel Panel (Committee), Fleet Air Arm Research Committee, Design Panel, Kite Balloon Sub-Committee (Panel), Scale Effect Panel, C.A.T.

Sub-Committee, Interference Panel, Air Cooling Panel, Whirling Arm Panel, RAF High Speed Wind Tunnel Panel, Free Flight Panel, High Altitude Sub-Committee, Helicopter Panel, Propeller Panel, Hypersonics Sub-Committee and the Powered Lift Committee.'[17]

- Evan James Williams FRS 'left his work at the University and joined the Instrument Department of the Royal Aircraft Establishment Farnborough (early 1940), working on submarine detection from the air. In March 1941 he joined the newly formed Operational Research Section at Coastal Command becoming Director of that organisation in December of 1941. During the last three years of the War the Naval Operational Research Department (NORD) at the Admiralty, contained the following senior staff: Sir Ralph Fowler FRS, E.J. Williams FRS, Sir Edward Bullard FRS, J.H.C. Whitehead FRS. Williams died on 29th September 1945 at the age of 42 in his parent's house at Brynawd Carmathenshire.'[18]

- Henry Charnock FRS 'was a trainee meteorologist in the Ministry of Supply.'[19]

- Henry Ellis Daniels FRS 'worked for the Ministry of Aircraft Production, mainly, it is thought, on missile problems.'[20]

- Otto Maass FRS 'became Director of Chemical Warfare and Smoke, National Defence Headquarters (Army) in 1941(in Canada), a post which he held until 1946. After the reorganisation of the Department of National Defence, he was appointed to the Staff of the Director General of Defence Research, and became Scientific Advisor to the Chief of General Staff (Army).'[21]

- Patrick Maynard Stuart Blackett FRS, Baron Blackett of Chelsea, 'served as Member of the Tizard Committee 1935–40, The Royal Aircraft Establishment Farnborough 1939–40, A.A. Command 1940–41, Coastal Command 1941–42, The Admiralty 1942–45.'[22]

- Sir Claude Dixon Gibb FRS 'left his post as Director-General of Weapons Production (in 1943) to become Director-General of Armoured Fighting Vehicles and later Chairman of the Tank Board. At the end of the War Gibb was offered posts with other large engineering concerns but chose to return to Parsons where he became Chairman and Managing Director in September 1945.'[23]

- Sir Ernest Marsden FRS 'was appointed Scientific Advisor to the New Zealand Fighting Services and was later given the rank of Lieutenant-Colonel (October 1939). He was Director of Scientific Development and Chairman of the Defence Scientific Advisory Committee throughout the War.'[24]

- Sir Frank Edward Smith FRS 'was appointed Director of Instrument Production by the Minister of Supply a post he held until 1942, in addition to the Chairmanship of the Ball and Roller Bearings Panel. He was also

Scientific Advisory Council Chairman from 1941 together with membership in a series of specialist Committees.'[25]

- Sir John Graham Kerr FRS 'was a Member of the Advisory Committee of Fishery Research from its beginning in 1919 and Chairman from 1942–49.'[26]
- Sir John Jacob Fox FRS 'joined the staff of the Director of Scientific Research in the Ministry of Supply, for part-time work, becoming Chemical Advisor to the Controller of Chemical Research.'[27]
- Sir Leonard Bairstow FRS 'served on many Committees both during the War and before/after including: Aeronautical Research Committee (ARC) Member (1921–45); Vice Chairman (1941–45); Aerodynamics Committee, Chairman (1931–46); Fleet Air Arm Research Committee, Member (1940–45), Chairman (1940–43); Engine Sub-Committee, Member (1926–45); High Altitude Sub-Committee, Member (1941–43), Chairman (1941–43).'[28]
- Sir William Scott Farren FRS 'became Deputy Director for Research and Development of Aircraft in the Directorate of Technical Development (1939). After the creation of the Ministry of Aircraft Production he became Director of Technical Development with direct responsibility to the Air Ministry. Soon afterwards he went to Farnborough as Director of the Royal Aircraft Establishment – the first to carry the new title (the old being Chief Superintendent).'[29]
- Sisar Kumar Mitra FRS 'became the first Chairman of the Board of Scientific and Industrial Research under the Department of Commerce of the Government of India.'[30]
- Thomas Percy Hilditch FRS 'was appointed Chairman of the Chemistry Technical Committee of the University of Liverpool Joint Recruiting Board. He served on the Food Investigations Board of the Department of Science and Industrial Research Investigation Board from 1934–46. Became Honorary Fire Observer in 1942 with his colleagues J. Proudman FRS, R.O. Griffith and R.A. Morton FRS, acted on behalf of the Fire Research Division of the Ministry of Home Security. In this regard he collected data on the origin and spread of fire and sent reports to the Fire Research Division in London.'[31]
- Thomas Smith FRS 'was appointed first Superintendent of the Light Division of the NPL in 1940.'[32]
- William Ernest Burcham FRS 'joined a small group under D.I. Dee working on aspects of air defence, first at the Royal Aircraft Establishment at Farnborough and then at the University College in Exeter, with the use of a small airfield at Haldon. He later accepted (1944) an invitation to join the British team working on atomic energy in Canada.'[33]
- William Jolly Duncan FRS 'joined the staff of the Royal Aircraft Establishment, Farnborough as head of the research section of the

Armament Department – particularly remembered for his contributions to the aerodynamics and control of winged torpedoes and gliding bombs. Took charge of the Air Defence Research Establishment in Exeter (June 1941), returning to the Royal Aircraft Establishment in 1942 as head of the Flight and Airborne Section.'[34]

4   In uniform, seconded

The individual work of the Fellows either in uniform or seconded to a military establishment will be covered in the relevant Chapter. Of interest here is the work of the Fellows which was of a broad scope affecting large numbers of people, or of Fellows in particularly unusual occupations.

For example the Medical Research Council instigated several key Committees whose remits were to consider the health and/or wellbeing of large numbers of military personnel across the Services. One Committee was established to investigate rations provided to the Services for all weather and terrain, namely the:

> Medical Research Council
> Military Personnel Research Committee
> (Rations) Sub-Committee[1]

Various meetings of this Committee were held during the War. The following were present for at least one of their meetings:

> Dr A.E. Carmichael (Chairman), Air Marshall Sir Harold Whittingham, Captain Bensley, Captain Finsen, Captain J.R. Poppen, Colonel A.A. Eagger, Colonel F.A.E. Crew, Dr Brian Roberts, Dr D.P. Cuthbertson, Dr E.B. Hughes, Sir George Lindor Brown FRS, Dr Ingram, Dr K.B. Turner, Dr L.H. Lampitt, Dr R.A. McCance FRS, Lieutenant Colonel M.H. Brown, Lieutenant Colonel Perrin Long, Lieutenant Colonel W.H. Griffin, Lord Falmouth, Major I.A. Anderson, Major Kreyberg, Major Pocock, Major T.R.B. Courtney, Mr A. King, Mr E. Bernard, Professor J.H. Burn FRS, Sir Henry Dale FRS, Squadron Leader T.R. Macrae, Wing Commander E.H. Anderson and Dr B.S. Platt (Secretary).

Meeting attendance was usually 7–8 people per meeting with some of the above individuals attending infrequently, depending on the topics under discussion.

The following subjects were discussed on at least one occasion:

- 2-day mess tin ration for use in temperate climates.
- Boiled sweets.
- Compact ration.
- Dehydrated foods.
- Dried fruit mixture.
- Field trial of the biscuit.
- Meat item – supplies from America of Cervelat sausage.
- Minced spiced ham product (known as SPAM).
- Modification for cold climate.
- Modification of 2-day mess tin ration for conditions of cold.
- Modifications of ration for periods longer than three days and for extremes of climate (hot and cold).
- Parachute and mobile troop ration.
- Ration for a 24-hour period.
- Report on field trial of W.D. mess tin ration.
- Report on the trial of compact ration on Canadian troops.
- Report on water and salt requirements of troops operating in very hot climates.
- Sugar-milk tablets.
- Supplies of salt to troops in the diet and by additions to drinking water.

Another Sub-Committee considered body armour and helmets:

Medical Research Council
Military Personnel Research Committee
(Sub-Committee on Body Armour and Helmets)[2]

Scientific personnel present during at least one of the Sub-Committee meetings:

Mr E. Rock Carling (Chairman), Professor J.D. Bernal FRS, Sir George Brown FRS, Dr C.S. Haldpike, Lord S. Zuckerman FRS , W.A. Carter, Dr E.A. Carmichael.

The 7th meeting, held on 13 April 1942 discussed the following:

- Helmet tests – Professor Zuckerman's report investigated the velocity of missiles of different weights and sizes necessary for penetrating skulls.
- Body armour for A.A. troops – the difficulty here was the need to obtain sufficiently light armour to move freely, yet still provide protection.

A meeting of the Sub-Committee held on 23January 1942 considered some of the issues of helmet design:

- That the chin strap was badly placed owing to faulty cut.
- That tests be carried out at the wind tunnel, N.P.L. to determine if some simple alteration to the leading edge could be designed to overcome the stream of air impinging on the eyes.
- That the inner band be lowered.
- That the rubber pad in the crown be attached by a screw and bolt.

At a meeting on 11 December 1941 there was an interesting discussion on the field trials of body armour covering:

- Chafing of the belly plate in the groin.
- Discomfort caused by sweating.
- Effects of the armour on a prismatic compass.
- Length of straps.
- Penetration tests.
- Protection of the genitals.
- The curvature of the breast place for comfort.
- The need for several sizes.

Many other issues were discussed – these are just some of the highlights.

At first sight the corresponding Habitability Sub-Committee does not appear to be populated by any Fellows. Actually there was one extremely distinguished Fellow, i.e. Surgeon Vice-Admiral Sir Sheldon Dudley FRS. Elected to Fellowship during the War (20 March 1941),[3] Sir Sheldon earned several medals and distinctions for his work on the spread of infectious diseases. He held the posts of Deputy Medical Director-General R.N. (1935–38), Medical Officer in Charge of the R.N. Hospital Chatham (1938–41) and was Medical Director-General of the Navy (1941–1945). He was awarded the Chadwick Gold Medal for the officer who had done most to promote the health of the Royal Navy.[4] There was no one more qualified to serve on this or indeed any of the other Committees. The subjects covered by this Committee included:

a) A comparison of the physiological effects of wearing anti-flash clothing and shorts in severe heat.
b) Carbon dioxide in magazines of HM Ships.
c) Energy output of 16in Mark III magazine crews.
d) Faulty perception caused by blank spells without signals during experiments in prolonged visual search.
e) Habitability and lighting of Fleet Air Arm workshops and hangers.

f)   Hot humid environment: its effect on the performance of a motor co-ordination test.

g)   Report on mock-up and cartridge handling trials in the 6in Q.F.S.A.R.P. 40 Mark XXV mounting.

h)   The ability to work in severe heat.

i)   The effect of the R.C.N. Seasickness Remedy on physical efficiency under Tropical Conditions.

j)   The effects of heat and high humidity on Pursuitmeter scores.[5]

In some cases there was the need to form various panels, for example the Panel on Turret Ventilation. Some of their reports covered topics such as:

1)   The consideration of possible alternatives to or modifications for measurement of air velocity in HM Ships.

2)   Living and working conditions among R.N. personnel in the Tropics.

3)   Climatic extremes influencing naval equipment.

In addition to the practical help given by the shore-based experts, there were also some new technologies to investigate:

24th July 1943
Dr T. Bedford
London School of Hygiene,
Kepple Street
London W.C.1.

Dear Dr Bedford,

Thank you for your letter dated 19th July concerning the types of insulating materials which have recently come under your notice. I was very interested to have this information and I have discussed it with the Section here which deals with this problem

The possibilities of 'Fiberglass' have been very much under consideration in recent days and experiments on the practical application of this type of insulating material have been planned.

'Isoflex' has been rejected by this department point owing to its inflammability a low melting point (the makers do not recommend its use in temperatures above 75 deg C).

We do not know very much about 'Foamglass' but a specimen is being obtained for trials.

My own feeling on the subject is that we are likely to get the best result with sprayed insulating material and for this reason it is considered that the early trials, anyway, should be carried out with sprayed asbestos. It is understood that D.N.O. is considering the possibility of carrying out such trials.

Yours sincerely,
Sims[6]

There was also a Committee looking at some of the general health issues of the day. This Committee was called the Army Medical Department Consultants' Committee. Such was the importance of the work this Committee undertook that the forty-fifth meeting was held on 9 December 1943. This was a remarkably large number of meetings, especially so given the regular attendees which included:

> Chairman – Lieutenant-General Sir Alexander Hood, Members Major-Generals A.B. Austin, D.C. Monro, C.M. Page, D.T. Richardson, A.W. Stott. Brigadiers W. Anderson, W. Rowley Bristow, Hugh Cairns, F.A.E. Crew FRS, Sir Stewart Duke-Elder FRS, H. Edwards, F.D. Howitt, G.W.B. James, R.M.B. MacKenna, D.B. McGrigor, R. Priest, G. Riddoch, H.A. Sandiford, J.V. Sinton, W.D.D. Small, F. Stammers, D.S. Stevenson, L.E.H. Whitby. In Attendance Dr J. Ferrebee, Colonels J.C. Kimbrough, L.C. Montgomery, J.A. MacFarlane, A.E. Porritt, R.M. Zollinger, Majors E.G. Colllins, A. Molls, S. Rowbottom. Secretary Lieutenant Colonel H.J. Bensted.[7]

This attendance list is from the forty-fifth meeting held on Thursday 9 December 1943. Even with such a distinguished attendee list, these were some unable to attend, i.e. Major-Generals A.G. Biggam and L.T. Poole, Brigadiers A.S. Daly, M.L. Formby, J.R. Rees, T.E. Osmond, H. Stobie, A. Hedley-Whyte, Colonels Elliott Cutler, F.H. van Nostrand and Dr Cuthbertson.

Francis Albert Eley Crew FRS was a Brigadier in the RAMC TA during World War Two. He was also Officer Commanding Military Hospital in Edinburgh Castle (1939–40), ADMS Edinburgh Area (1940–42), Director of Medical Research WO (1942–46) as well as Buchanan Professor of Animal Genetics (1928–44), Director of the Institute of Animal Genetics (1921–44) (University of Edinburgh) and Professor of Public Health and Social Medicine Edinburgh University (1944–45). He served in India and France in World War One with the 6th Battalion The Devonshire Regiment and No 3 FA, RAMC Guards Division.[8]

Sir Stewart Duke-Elder FRS served in the Army during World War Two as a Brigadier in the RAMC. He was appointed Consulting Ophthalmic Surgeon to the Army in 1946 (until 1961).

It is possible that Brigadier J.V. Sinton mentioned above should have read Brigadier J.A. Sinton VC FRS. The latter Brigadier is the only individual to hold both the Victoria Cross and Fellowship of the Royal Society. Formerly retired prior to World War Two, Brigadier J.A. Sinton VC FRS was re-employed by the army for the whole of the duration of World War Two, spending his time as a consultant to the Army on malaria and entomology.

The minutes of the forty-fifth meeting, of which the above were the attendees, considered several issues, typical of some of their other meetings, and included discussions on:

a) Insufflators for sulphonamide and penicillin powder.
b) Mechanical respirators.
c) Penicillin.
d) Pre-sterilised dressings.
e) Rehabilitation of officers.
f) Sciatic pain.
g) Special diseases of and injuries to blood vessels.
h) The need for a camera for the Maxillo-facial units.
i) Training of surgeons in field surgery.

Two Sub-Committees were mentioned in the minutes, one called the Medical Sub-Committee and the other the Surgical Sub-Committee.

One might image that this was one of the key medical Committees given the illustrious attendees. Perhaps the following might help to confirm this assumption. An extract from the forty-third meeting, held on 14 October 1943 contained an article on penicillin.

> The Chairman announced that the Canadian offer of penicillin for the British Army had been passed to the Ministry of Supply and explained that this offer would be included in the general allocation plan of supplies controlled by the Ministry of Supply.

The report mentioned the quantities available for experiments:

> The Penicillin Clinical Trials Committee at a meeting on 12th October allotted to the Army for research purposes 110,200,000 Oxford units to be allocated as follows:
> a) 20,000,000 units for trials in gas gangrene.
> b) 20,000,000 units for trials in wounds of chest.
> c) 50,000,000 units for trials in compound fractures.
> d) 20,000,000 units for trials in sulphonamide-resistant meningitis.
> e) 200,000 units for trials in chronic supp. Otitis Medis.
> In addition to this allotment for clinical trials, the Ministry of Supply had

been asked for immediate delivery of 50,000,000 units for the treatment of conditions already known to respond to penicillin. It was hoped that this would be available in the course of the next two months.

Previous minutes also mention the use of benzedrine in War operations thus:

> benzedrine sulphate tablets may be taken by mouth by fatigued subjects to reduce the symptoms of fatigue and particularly the desire for sleep.[9]

Of course the clinical trials of penicillin used culture grown material as the structure and manufacture of synthetic penicillin required the structure to be elucidated. Additionally a proven method of manufacture was needed, more of which in Chapter 9.

SOE/SBS/SAS

The increasingly unsettled times of the mid to late 1930s prompted the Government to form three separate organisations as part of the UK preparations for War (in 1938):

- Electra House (a semi-secret propaganda section of the Foreign Office).
- MI R (a research branch of the War Office).
- Section D (the Sabotage Branch of MI6).[10]

Soon after his election to Prime Minister, Sir Winston Churchill authorised the amalgamation of these three organisations, forming the Special Operations Executive (SOE). The National Archives describe the SOE's role as:

> To promote sabotage and subversion in enemy occupied territory and to establish a nucleus of trained men tasked with assisting indigenous resistance groups.

Initially run by Sir Frank Nelson its first Chief Executive Officer, the SOE was organised into three departments, namely:
a)   SO 1 (propaganda).
b)   SO 2 (active operations – this branch was subsequently split into groups dealing with geographical areas of operation).
c)   SO 3 (planning).

SO 1 was transferred to a newly created Political Warfare Executive in August 1941. Very few records of the SOE are known to have survived, however, some of the documents relating to several of its operatives are available in the National Archives. Figure 4.1 details just a few of these resourceful and brave individuals who lost their lives.

Name	Cause of death	National Archive Reference
Andre Burguiere	French agent, died in swimming accident while behind enemy lines.	HS 9/237/7
Andre Dubois	French W/T operator recruited in the field, arrested in November 1943 and later killed.	HS 9/451/6
Charles Gaskell	Liaison officer with Russian NKVD; killed in plane crash, November 1943.	HS 9/566/8
Edward Bisset	Arms instructor dropped into France, accidentally killed September 1944.	HS 9/158/2
Emile Garry	Lieutenant in Suttill's Prosper circuit; arrested September 1943, killed at Buchenwald, 1944.	HS 9/566/2
Howard Burgess	Instructor; died of natural causes, June 1942.	HS 9/237/2
John Ali Mackintosh	Worked under commercial cover in Turkey, died of typhoid in January 1945.	HS 9/965/8
Olive Burgess	Secretary, SOE India; killed in plane crash, Trincomalee, October 1945.	HS 9/237/3
Paul Halley	Canadian linguist, operational officer, SOE India, killed in car crash, December 1944.	HS 9/648/1
Philippe Duclos	French agent dropped to a compromised circuit in February 1944, arrested on landing, killed by September 1944.	HS 9/453/7
Stanley Mackintosh	36 military mission, India, died of typhoid, June 1945.	HS 9/965/9
Stephen Leake	Liaison officer in Albania, killed in enemy air raid.	HS 9/900/1
Thomas Handley	W/t operator, Greece, captured and killed after attack on Corinth Canal.	HS 9/653/6

Figure 4.1 SOE Operatives who lost their lives.

Figure 4.2 is a partial list of known SOE operatives known to have survived the War.

Name	Activity	National Archive Reference
Amedee Maingard		HS 9/976/9
Boris Hembry	Saboteur behind enemy lines, south east Asia; walked out of the jungle.	HS 9/691/1
Clara Holmes	Intelligence agent in pre-war Austria, worked propaganda into Germany and Austria until the fall of France and then worked for SOE Austrian section.	HS 9/733/4
Clement Marc Jumeau		HS 9/815/4
Edward Halton	Wireless operator, India.	HS 9/648/7
Edward Mayer	Temporary officer at SOE HQ, posted out May 1943.	HS 9/1011/2
Francis Suttill		HS 9/1430/6
Fridtof Normann	Norwegian agent, sent on a mission, November 1942, missing presumed killed.	HS 9/1110/8
George Begich	Canadian Croat, SOE translator.	HS 9/114/9
George Sutton	Courier, worked to set up secret arms dumps in the UK against possible German invasion.	HS 9/1430/3
Gerald Sutton	Engineer, worked on design of SOE secret weaponry.	HS 9/1430/4
Gilbert Norman		HS 9/1110/5
Henri Hubert Gaillot	Belgian agent planned sabotage in Brittany but arrested in February 1944.	HS 9/554/1
Henry Byrne	Signals officer.	HS 9/251/5
James Andrew Mayer	Madagascan w/t operator, dropped into France February 1944, arrested in May and killed, September 1944.	HS 9/1011/4
John Beevor	Directed SOE infiltration to Yugoslavia; directed Lisbon mission.	HS 9/114/6
John Kenneth MacAlister		HS 9/954/2
Julius Hanau		HS 9/653/2
Luigi Galgani	Italian anti-Fascist; transferred to PWE India Mission, 1942.	HS 9/554/8
Max Hymans	Resistance leader, former French Under-Secretary of State.	HS 9/773/3
Noor Inayat Khan		HS 9/836/5
Norah Galbraith	FANY with SOE headquarters.	HS 9/554/4
Odette Hallowes	Remains the only woman to have received the George Cross whilst alive.	HS 9/648/4

Name	Activity	National Archive Reference
Pierre Culioli	French resistance fighter, imprisoned at Buchenwald after capture in 1943, but survived the War.	HS 9/379/8
Rene Burgraeve	Belgian postmaster, twice landed in Belgium to collect intelligence.	HS 9/237/6
Robert Byerly	American dropped into France and arrested on landing, February 1944.	HS 9/251/1
Stanley Casson	Liaison officer with the Greeks.	HS 9/278/5
Vera Atkins		HS 9/59/2
Wallace Leaper	Liaison officer.	HS 9/900/2
William Arthur Knox		HS 9/852/1
William Hancock	Engineer, SOE Middle East.	HS 9/653/4
William Mailer	Liaison officer, Malaya.	HS 9/976/6
Yolande Beekman		HS 9/114/2

Figure 4.2 Partial list of known SOE operatives known to have survived the War.

Figures 4.1 and 4.2 do not contain the names of any Fellows and so at first glance it may seem curious that these SOE operatives have been mentioned here. However, there were some Fellows who served in the SOE, and may have known or supported those listed in Figures 4.1 and 4.2.

Sir Ernest Gordon Cox FRS joined the Territorial Army in 1936 and was an officer in the Birmingham University OTC.[11] He had also led an advisory group on explosives for the Ministry of Supply. Cox was recruited by D.M. Newitt FRS Director of Scientific Research for the Inter-Services Research Bureau (ISRB was the cover name for the Special Operations Executive (SOE)) to be a Senior Officer in charge of the ISRB station at The Frythe, Welwyn Herts (February 1942). In the summer of 1944 Cox went to France as Technical Staff Officer in the 21st Army Group HQ with the rank of Lieutenant Colonel.

One of his tasks in his meetings with the underground was the investigation of V-2 rocket sites. Cox had close ties with the Belgian Resistance Group G.

Clement Henry Bamford FRS worked for the Royal Aircraft Establishment, Farnborough, and subsequently for the Special Operations Executive. The nature of his work, although not recorded in detail, included such things as the mechanism of action of incendiary agents and combustion processes.[12]

Bamford's *Who was Who* entry provides the year he joined the SOE which was 1941.[13]

Figure 4.3 Sir Ernest Gordon Cox. (©
Godfrey Argent Studio. Reproduced
with permission)

Figure 4.4 Clement Henry Bamford.
(© Godfrey Argent Studio. Reproduced
with permission)

Sir Alan Herries Wilson FRS was recruited into the radio communications laboratory of the SOE (September 1941) which had the task of developing communications between Britain and the resistance groups ... he joined the Tube Alloys project in 1943 having completed his task.[14]

Douglas Hugh Everett FRS studied chemistry at Balliol College Oxford as a Ramsay Memorial Fellow. He was awarded a DPhil in 1942. Recruited into the SOE as a senior scientific officer at The Frythe, Hertfordshire, commissioned Major (1944).[15]

George Patrick John Rushworth Jellicoe FRS, 2nd Earl Jellicoe was part of Colonel Robert Laycock's Layforce, and then with David Stirling and others in the early Special Air Service (SAS) and SBS, of which he was the first Commander. He was promoted to Lieutenant-Colonel aged 25 years. He was three times mentioned in despatches, wounded, awarded the Distinguished Service Order (DSO) for his part in the raid on Haraklion airfield in Crete. He was captured in the Dodecanese, and later escaped. He was awarded the DSO in 1942, a Military Cross in 1944, the Legion d'Honneur, the Croix de Guerre and the Greek War Cross. His civil honours were PC 1963; KBE 1986.[16]

Others in uniform

Arguably Louis Francis Albert Victor Nicholas Mountbatten FRS, 1st Earl Mountbatten of Burma was the most high profile Fellow in uniform during World War Two. Elected under Statute 12 of the Royal Society regulations on 19 May 1966 he is included here as he was associated with the Royal Society at some point during his lifetime.

Earl Mountbatten held many civil honours, i.e. MVO (1920), KCVO (1922), GCVO (1937), DSO (1941), CB (1943), KCB (1945), KG (1946), PC (1947), GCSI (1947), GCIE (1947), GCB (1955) and OM (1965). He was ennobled as Viscount Mountbatten of Burma 1946, Earl in 1947 and Baron Romsey also in 1947.[17]

His World War Two activities included:

> Commanded 5th Destroyer Flotilla in HMS *Kelly* 1939–41 (dispatched twice), HMS *Illustrious* 1941, Commodore Combined Operations 1941–42, Chief of Combined Operations and Member of the British Chiefs of Staff Committee 1942–43, Acting Vice-Admiral 1942, SACSEA 1943–46 (Acting Admiral 1943), Supreme Allied Commander South East Asia 1943–46, Rear Admiral 1946.[18]

He was awarded the following for War Service:

> Legion of Merit (1943), DSM (1945, US), Greek Military Cross (1941, Crete) , Grand Cross of the Order of George I (1946, Greece), Special Grand Cordon of the Cloud and Banner (1945, China), Grand Cross of the Legion of Honour and Croix de Guerre (1946, France), Grand Cross of the Star of Nepal (1946), Order of the White Elephant of Siam (1946), Order of the Lion of the Netherlands (1947).[19]

He was awarded the following but not for War Service:

> KStJ (1943), Grand Cross of Isabella Catolica (1922, Spain), Crown of Rumania (1924), Star of Rumania (1937), Military Order of Avis (1951, Portugal), The Seraphim (1952, Sweden), Agga Maha Thiri Thudhamma (1956, Burma), Grand Cross, Order of Dannebrog (1962, Denmark), Grand Cross of the Order of the Seal of Solomon of Ethiopia (1965).

Some of the other fighting Fellows who served in the armed forces are listed in Figure 4.5.

Name	Posting
Billingham (Rupert Everett)	Went to Portsmouth to receive training as a Temporary Acting Probationary Sub-Lieutenant in the Royal Navy Volunteer Reserve. Sent to Greenock on the Clyde as a shore-based Radar Officer (early 1943), helped to service radar equipment fitted to anti-submarine vessels of the Clyde Escort Force.[20]
Blin-Stoyle (Roger John)	Conscripted in 1943 and assigned to the Royal Corps of Signals. Promoted to the rank of 'local unpaid lance-corporal' and recognised as potential officer material.[21]
Cain (Arthur James)	Served as a Second Lieutenant in the Royal Army Ordnance Corps, and later transferred to the Royal Electrical and Mechanical Engineers. Responsible for anti-aircraft radar, protecting the ships of the fleet in the Orkney Islands.[22]
Fuchs (Sir Vivian Ernest)	Became Adjutant to the Second Battalion of the Cambridgeshire Regiment, later joining Brigade HQ as Transport Officer. Later in the War he qualified and was posted to Second Army HQ to work in Civil Affairs. His unit left London for Plymouth to prepare for the invasion of France. The day after D-day, his group embarked, taking three days to reach Gold Beach.[23]
Gray (Edward George)	Served as an Able Seaman from 1942 to 1945.[24]
Hammersley (John Michael)	Commissioned as a second lieutenant in the spring of 1941 and posted to an anti-aircraft gun site defending an armament factory near Worsham. Took the unusual step of telephoning divisional headquarters and as a result was selected to train to become an I.F.C.[25]
Hewitt (John Theodore)	Received a commission as 'Major, General List attached to the Royal Engineers' and with H.R. Le Sueur acquired a uniform and started for the Dardinelles.[26]
Hirst (John Malcolm)	At his own request he was transferred to Coastal Forces where his duties included mine-laying, mine-sweeping, convoy escort and clandestine coastal duties.[27]
Hogg (Quintin Mcgarel, Lord Hailsham of St Marylebone)	Sent out to the Western Desert, where he was wounded. There followed a spell in Beirut, where he was given a liaison job between the Army and the Free French.[28]
Katz (Sir Bernard)	First military posting (Oct. 1942 to Mar. 1943) was with a radar unit on Goodenough Island (New Guinea). Firstly as a radar pilot officer, and later as a Flight Lieutenant, he was in charge of about 20 men running a movable radar unit. His second posting (mid-1943 to autumn 1945) was at the Radio-Physics Laboratory at Sydney University, where he was involved in the development of the radar transponder.[29]

Name	Posting
Matthews (Richard Ellis Ford)	Appointed to the DSIR staff as Assistant Mycrologist (Nov. 1941) but the following month he was called up for active service. Posted to the 7th Anti-Tank Regiment of the 2nd New Zealand Division (1943) and took part in the whole campaign of the 8th Army in Italy from 1943 to 1945.[30]
McLaren (Digby Johns)	Joined the Royal Artillery as a gunner (1940). Gained his commission and posted to Iraq. Fought across that country with his battery, then through Italy with the Eighth Army in 1943–45. Finished the War as a Captain.[31]
Neumann (Bernard Herman)	Initially interred. He was released into the British Army after several months. Trained in Yorkshire and then transferred to the south of England. Served in the Pioneer Corps until 1943 and was then allowed to volunteer for combatant service which he did in the Royal Artillery, moving from there into the Artillery Survey Corps.[32]
Norrish (Ronald George Wreyford)	Added somewhat to his age, joined the Royal Field Artillery and served as a Lieutenant, first in Ireland and then on the Western Front.[33]
Northcott (Douglas Geoffrey)	Joined the Royal Artillery as a volunteer (Nov. 1939). During Northcott's first overseas posting in India, he contracted the first of a number of illnesses. Later in the War a Royal Army Medical Corps doctor operated on him in a makeshift hospital for what turned out to be peritonitis.[34]
Pereira (Sir Herbert Charles)	Appointed a Company Commander and later, Captain in the RE. Formal notification of the award of his degree was delivered by a motorcycle dispatch rider. Travelled all over the Middle East to reconnoitre potentially defensive positions. Pereira was Mentioned in Dispatches.[35]
Rees (Hubert)	Joined 75th squadron (RNZAF) in 1944 to pilot Lancaster bombers. At the end of his first tour of 30 operations he was awarded the DFC. Soon after commencing his second tour he was shot down over Homberg Germany and following capture was sent to Stalag Luft 1 on the Baltic Coast.[36]
Spear (Walter Eric)	Joined the Pioneer Corps in 1940 and later transferred to the Royal Artillery. Demobilised with the rank of Bombardier (1946).[37]
Wareing (Philip Frank)	Joined the REME in Dec. 1941 and was mainly involved in RADAR for coastal artillery. He had a number of postings and ended the War as Captain in Trinidad.[38]

Figure 4.5 Fighting Fellows.

Of course there were some Fellows in uniform who undertook technical duties.

1) Bertram Neville Brockhouse FRS 'enlisted in the Royal Canadian Navy and spent some months at sea. However he spent most of the War on land servicing ASDIC (Anti-Submarine Detection Investigation Committee) equipment. On graduating from a six-month course in Electrical Engineering at the Nova Scotia Technical College, he became a Sub-Lieutenant and was assigned to the test facilities of the National Research Council of Canada in Ottawa. He was subsequently awarded the Nobel Prize (Physics) 1994.'[39]

2) Edward Norton Lorenz FRS 'signed up for a course as an aviation cadet in the then Army Air Corps (now the Air Force). Five from this course, Lorenz included, were chosen to stay on as teachers for the new course. He was later posted to Saipan and Okinawa in the western Pacific Ocean (1944), where his principle job was to forecast winds and temperatures in the upper troposphere.'[40]

3) George Hugh Henderson FRS 'in 1943 he assumed full charge of HM Canadian Naval Research Establishment and also became Operational Research Adviser to the Commander-in-Chief Canadian North-West Atlantic.'[41]

4) Harry Elliot FRS 'joined Coastal Command. After initial training in Uxbridge and officer training in Cranwell, Elliot was posted to Devon as a signals officer.'[42]

5) John Heslop-Harrison FRS 'after a tour of duty at a remote gun emplacement, he had a roving inspectorate of all radar establishments having first been commissioned into the RAOC and then transferred to the REME.'[43]

6) John Stanley Sawyer FRS 'took charge of the meteorological office at RAF Thorney Island near Portsmouth(Sept 1939), providing forecasts for aircraft engaged in coastal reconnaissance and attacks on German shipping. Like most meteorologists, he was put into uniform and commissioned as a Flight Lieutenant in the RAF Volunteer Reserve. He was subsequently posted to Delhi as deputy to the Chief Meteorological Officer at Air Command Headquarters.'[44]

7) Laurence Rickard Wager FRS 'shortly after the War started Wager was commissioned in the Royal Air Force and entered the photographic reconnaissance section. Wager served in the United Kingdom, in the Middle East and for a short period in Arctic Russia. He was mentioned in dispatches. In 1944 he was released from the RAF to take the Chair of Geology at the University of Durham.'[45]

8) Lord Edward Arthur Alexander Shackleton KG FRS 'worked for
    the Ministry of Information before joining the RAF in 1940. He
    was stationed with Coastal Command as a station intelligence officer
    and anti-U Boat planner at St Eval in Cornwall. In 1943 he moved to
    Coastal Command HQ in Northwood.'[46]

10) Richard Darwin Keynes FRS 'was transferred to the Admiralty Signals
    Establishment at Witley in Surrey (1942) where he was concerned with
    improving radar control of naval anti-aircraft guns.'[47]

11) Sir John Augustine Edgell FRS 'served for fifty one years in the Royal
    Navy, 43 of them in the Surveying Service. He was promoted to Rear
    Admiral in 1935 and Vice Admiral in 1938. He was then placed on the
    reserve list, but remained in his post as Hydrographer of the Navy till
    the end of the War.'[48]

12) William Bernard Robinson King FRS 'called up immediately War was
    declared from the Army Officers Emergency Reserve as a geologist
    with the initial rank of Major to the Engineer in Chief BEF (France).
    King was attached to Northern Command for a year and then from
    1941 to 1943, to GHQ Home Forces which was to become for inva-
    sion purposes 21 Army Group. He worked with the planners on the
    invasion of Normandy. He initiated a whole series of enquiries into
    water supply, trafficability, availability of constructional materials and
    operational airfield sites on the French mainland and on the nature
    and distribution of beach materials on those shores where the invasion
    army might land. Indeed it had been publically stated by General Sir
    Drummond Inglis that Cotentin was selected as the original bridge-
    head because of the military necessity of Cherbourg as a port, and it
    was only when King refused to 'produce' a suitable geology for a large
    number of fighter airfields, but was able to do so in Normandy, that the
    invasion site was changed and the idea of the artificial Mulberry ports
    took shape.'[49]

13) William Reginald Stephen Garton FRS 'volunteered and served in
    the Royal Air Force, being posted as a meteorological officer in North
    Africa. Meteorological officers were few in number, and they were
    regarded as important and were therefore well protected.'[50]

As one might imagine, some of the Fellows with medical qualifications
made invaluable contributions to the Royal Army Medical Corps (RAMC)
or its equivalent.

Name	Posting
Colebrook (Leonard)	Appointed Colonel in the RAMC and Bacterial Consultant to the British Expeditionary Force, then Director of the Burns Unit in Glasgow Royal Infirmary (1942). Appointed Director of the Burns Unit at the Birmingham Accident Hospital until his retirement in 1948.[51]
Dacie (Sir John Vivian)	Was moved to the Central Pathology Laboratory at Epsom (Jan 1940). Transferred to the RAMC (early 1943). His unit landed in Normandy six days after D-day and acted as a Casualty Clearing Station. After the advance into The Netherlands he spent some time studying blood loss in wounded soldiers and was then posted to Italy.[52]
Doll (Sir William Richard Shaboe)	He volunteered for the Royal Army Medical Corps, with which he served throughout the War. Doll treated and helped evacuate many wounded through the retreat to Dunkirk, despite sustained shelling and air attacks.[53]
Fairley (Neil Hamilton)	Colonel in the Australian Army Medical Service. Joined the Headquarters of the Australian Forces in Cairo as Consulting Physician (1940). Promoted to Brigadier and became Director of Medicine in the Australian Military Forces. Soon afterwards he was appointed Chairman of the Combined Advisory Committee on Tropical Medicine, South Pacific Area and in this capacity was directly responsible to General MacArthur.[54]
Morgan (Walter Thomas James)	Met Dr John Loutit (FRS), Director of the South London Blood Transfusion Depot at Sutton and started a program at the Lister Institute to study the chemical nature of ABO blood groups.[55]
Rosenheim (Max Leonard), Baron Rosenheim of Camden	Joined the RAMC in 1941. After a preliminary assignment in Belfast, he became Officer in Charge, Medical Division in various countries in the Middle East, North Africa and Europe, ending his Army service as consulting physician to the Allied Forces, S.E. Asia.[56]

Figure 4.6 Fellows who served in the Royal Army Medical Corps (RAMC) or its equivalent.

Of course not all scientists and/or engineers were called into the Services, some were seconded.

1) Alfred John Sutton Pippard FRS 'joined the Research and Experiments section of the Civil Defence Research Committee located at Princes Risborough, whilst also teaching young engineers and scientists work at Imperial College.'[57]

2) Charles Henry Brian Priestley FRS 'was appointed to the Meteorological Office as a Technical Officer in April 1939. He was chosen to join a small research group in micrometrology led by Sir Graham Sutton FRS at the Chemical Defence Experimental Station at Porton Down in Wiltshire. During his two years at Porton Down Priestley worked with F.A. Pasquill FRS. Unexpectedly, he was recalled to work in the newly formed Upper Air Analysis and Forecast Section (the Upper Air Unit) at the Meteorological Office in Dunstable (1943) and participated in the successful D-Day weather forecast. In September 1944 he was promoted to Senior Meteorologist and officer in charge of the Synoptics Section of the Upper Air Unit.'[58]

3) Clifford Hiley Mortimer FRS 'was called up as a civilian scientist at the Admiralty Mine Department. Later he went on to other aspects related to waves and water movement, for example in relation to harbour defence and the D-Day landings ... He finally found himself in 'Group W' (for waves) of engineer and scientists. The same group contained Francis Crick FRS, Sir George Deacon FRS, Howard Penman FRS, Michael Longuet-Higgins FRS and Sir Anthony Laughton FRS. Group W later provided the nucleus of the new National Institute of Oceanography.'[59]

4) Edward Foyle Collingwood FRS 'became a naval scientist in the Mine Sweeping Division with Sir Edward Bullard FRS. Unlike other naval scientists he 'got himself into uniform' early in the War. He became Temporary Lieutenant RNVR in 1940, Lieutenant-Commander in 1942, Commander in 1942, Captain in 1944. One of a very few captains in the RNVR. He served as Director of Scientific Research with the Admiralty delegation to Washington in 1942, as officer in charge of the Sweeping Division in 1943, Chief Scientist Admiralty Mine Design Department 1943–45 and also one of a delegation to Moscow on a special scientific mission.'[60]

5) Edward Neville Da Costa Andrade FRS 'continued in office at University College but became increasingly occupied with scientific advisory work for the Ministry of Supply. He was particularly involved in scrutinizing ideas for inventions that were then flooding into the Government.'[61]

6) Emmanuel Fauré-Fremiet ForMemRS 'became involved with an organisation for studying chemical warfare, and as a member of the Council of the Centre National de la Recherche Scientifique was on the Committee for the study of the problems of wound healing.'[62]

7) Francis Albert Eley Crew FRS 'shortly after his appointment as Director of Biological Research at the War Office (with the rank of Brigadier) he was asked to undertake writing the official Army Medical History of the War.'[63]

8) Harold Haydon Storey FRS 'was seconded to the East African Supplies Board as Secretary to an Industrial Advisory Board in Nairobi (1941).'[64]

9) Horace Newton Barber FRS 'attached to the Telecommunications Research Establishment and helped to introduce and incorporate in the working of the Air Force the succession of radar inventions.'[65]

10) Ivan de Burgh Daly FRS 'went down to Farnborough (autumn, 1939) to see what could be done in the Edinburgh laboratory to assist the Royal Air Force. As a result a decompression chamber was set up and a lot of experiments were done on small animals with particular reference to bends over 35,000ft. At the same time, a great deal of research was done by much the same team, on the effect of phosgene gas on the lungs. In January 1943, at the invitation of Sir Edward Mellanby FRS, Secretary of the Medical Research Council, Daly was temporarily seconded by the University of Edinburgh to Direct the MRC Physiological Laboratory attached to the Gunnery Wing of the Army Fighting Vehicles Training School.'[66]

11) John Barker FRS 'spent several years on the staff of Sir Jack Drummond FRS, Scientific Advisor to the Ministry of Supply, acting as scientific advisor to the Director of the Ministry's Dehydration Division.'[67]

12) Maurice Neville Hill FRS (son of A.V. Hill FRS) 'applied for a transfer to the Sweeping Division of the Mine Design Department located in Edinburgh. He arrived in Edinburgh in September 1941 and gradually took on wider responsibilities, becoming Group Leader in 1944. His main interest was in the counter measures to the German acoustic homing torpedo, this involved an investigation of the properties of the weapon and the devising of sound sources to mislead it.'[68]

13) Richard Whiddington FRS 'spent the whole War "on loan" to the government. For a short time he served on an RAF officers' selection board, with the rank of Wing Commander. Later he was with the Admiralty Scientific Service and was involved in furthering the development of radar equipment for use by the Navy. Finally he was appointed Deputy Director of Scientific Research in the Ministry of Supply.'[69]

14) Ronald Percy Bell FRS 'worked half-time in the Scandinavian Section of the Royal Institute of International Affairs (Chatham House), its Foreign Research and Press Service and also in its successor, the Foreign Office Research Department. Their main functions "were to scan the foreign press, to compile press cuttings, and to write periodical reports about events in enemy, occupied, and neutral countries." Bell commented … "We also occasionally interviewed people who had escaped from enemy controlled territory" … One of the escapees that Bell helped to debrief was Niels Bohr ForMemRS.'[70]

15) Samuel Phillips Bedson FRS 'seconded to the Emergency Medical Service and was pathologist to sectors I and II of region V (Billericay, Essex).'[71]

16) Sir Clifford Charles Butler FRS 'graduated in 1942 with a first-class BSc (Special). He stayed on as a demonstrator, his National Service taking the form of teaching radio as part of the State scheme to produce radar physicists, and acted part-time as a physicist at the Royal Berkshire Hospital.'[72]

17) Sir Harry Raymond Pitt FRS 'moved to Aberdeen University (1939) as an assistant lecturer. He later moved to London for work at the Air Ministry and the Ministry of Aircraft Production (1942) for three years, under H.E. Daniels FRS. In 1945, with the War ended, Pitt was appointed Professor of Mathematics at Queen's University, Belfast, at the age of 31 years.'[73]

18) Sir Paul Gordon Fildes FRS 'seconded to the Experimental Research Station at Porton (1939–45) investigating the exploitation of bacteria in warfare.'[74]

19) Sir William Hunter McCrea FRS 'moved to Queen's University, Belfast, as Professor of Mathematics (1936). He was given leave from Belfast while doing operational research in the Admiralty in the team led by Lord Patrick M. S. Blackett FRS.'[75]

20) Sir William Kershaw Slater FRS 'Scientific Advisor to the Ministry of Agriculture (from 1942) and secretary of the Agricultural Improvement Council, appointed Senior Advisor to the Ministry of Agriculture in 1944.'[76]

21) Sydney Chapman FRS 'worked with G.I. Finch on incendiary bomb problems (1942), during 1943–45 he worked at the War Office on problems of military operational research.'[77]

22) Theodore von Kármán ForMemRS 'was officially a consultant to the US Army Air Corps and later to the US Air Force (from 1939). One of the first results of his efforts was a change in policy, whereby the Air Corps was authorised to conduct its own research and development, and he became the chief consultant in the design of the 40,000 horse-power

20ft wind tunnel at Wright Field, Dayton, Ohio. Towards the end of the War (1944), General Arnold asked von Kármán to form and act as Chairman of a Scientific Advisory Group to study the use of science in warfare in Europe and Japan and to interpret the significance of new technological developments for the future of the US Air Force.'[78]

23) Thomas MacFarland Cherry FRS 'applied his mathematical analysis skills to the detection of aircraft by radar, pressure and temperature generated in a film of nitroglycerine when hit by a hammer etc.'[79]

There are many Fellows, some mentioned above, who gave their energy and time to the war effort, as did, of course, the rest of the country. Two Fellows are selected for discussion here for some of the more unusual aspects of their war-work.

• Joseph Arthur Colin Nicol FRS. 'Whilst at a holding camp in Farnham Nicol volunteered for a unit using homing pigeons, thinking there would be some biology involved. He spent several interesting months managing a pigeon troop, watching keen pigeon fanciers training birds for mobile warfare and employment at night. Being the scientist that he was he started to observe their behaviour in an attempt to establish how they navigated. His experiences led him to conclude that vision was the main factor in homing ability. The birds learnt the appearance of the countryside around their home loft, or anywhere they were fed and watered. Attempts to get the pigeons to fly at night showed that they took much longer to learn and longer to return. He later followed the invading troops into France, landing at Courcelles and moving on daily after the retreating Germans. Crossing into The Netherlands he spent the winter near Nijmegen, came under infantry attack at Otterloo while moving to relieve Rotterdam, ending up at Groningen.'[80]

Sir Godfrey Newbold Hounsfield FRS must have been an interesting individual. A comment in his Royal Society Biographical Memoir states:

After the War, Air Vice-Marshal Cassidy arranged for a grant to be given to Hounsfield to enable him to attend Faraday House Electrical Engineering College in London, where he received a diploma. This was to be Hounsfield's highest formal educational qualification.[81]

Arguably this statement on its own is not of particular interest, unless of course one were to look at the website which lists Nobel Laureates, i.e.

The Nobel Prize in Physiology or Medicine 1979 was awarded jointly to Allan M. Cormack and Godfrey N. Hounsfield 'for the development of computer assisted tomography'.[82]

In agreement with the statement of the Biographical Memoir, no formal university qualifications are mentioned in the Royal Society Fellows database.[83] His entry in that database, however, does list his post War activities which led to his Nobel Prize:

- Research staff Electric and Musical Industries (EMI) Hayes (1951).
- Moved to EMI Central Research Laboratories where he developed the EMI brain scanner – Head of Medical Systems Section, Thorn EMI Central Research Laboratories (1972–1976).
- Chief Staff Scientist (1976–1977).
- Senior Staff Scientist (1977–1985).
- Consultant to Laboratories (1986–2004).

Hounsfield's achievements may be unique in the annals of the Nobel Foundation records – at least for the science Prizes. Certainly there are no known Chemistry Nobel Laureates without a formal education to at least first degree level.[84]

But what of Hounsfield's war-work? His biographical Memoir comments:

> At the outbreak of World War Two, Hounsfield joined the Royal Air Force (RAF) as a volunteer reservist. This gave him the opportunity to study the books that the RAF provided for radio mechanics. He took a course in radio, passed a trade test and was taken on as a radar mechanic instructor. His first posting was to the Royal College of Science in South Kensington – then occupied by the RAF – and, later, he was based at Cranwell Radar School. He rose to the rank of Corporal. In his spare time at Cranwell, Hounsfield sat and passed the City and Guilds examination in radio communications. He also built a large-screen oscillo-scope for use in teaching, for which he was awarded a certificate of merit.[85]

For such a high profile scientist, little has so far been published concerning Sir Godfrey's character. The author recently had the privilege of meeting a fellow radar instructor from Sir Godfrey's RAF Cranwell days. Not only did they share some of the teaching duties they were also billeted in the same hut. Seventy years since they worked together in No 8 Radar School at Cranwell, the memories of Sir Godfrey are still as fresh as they were.

Sir Godfrey is described as a character. He was the one man who everyone turned to for help with any aspect of radar or indeed any other engineering problem. Encouraged to wear civilian clothes for VE night celebrations, Sir

Godfrey designed and built a piece of head wear that flashed every time he took a step. He was fascinated by any form of engineering and on VJ night spent time making an old nearby steam roller serviceable. He would often skip meals to think of solutions to problems presented to him and on one occasion used his pyjama cord as a makeshift fuse. He is still described with affection as someone with a unique insight always interested in helping anyone with problems. These traits were also used by him in his professional work as he invented a diagnostic system for checking radar equipment without any prompting from his chain of command. He ended the War as a Flight Sergeant.

5   Reserved Occupations

There were many and varied justifiable reserved occupations. As one might imagine the Fellows were as well represented in all aspects of these occupations as were the rest of the civilian population. Some of the Fellows, typically those too old to fight, took on greater teaching and administrative roles within our academic institutes. Others carried on their academic posts but spent the majority of their time as consultants for various wartime activities. A further group of Fellows were actually employed by Government Institutes or industrial concerns, the output from which was vital to the war effort. In some cases these individuals joined their respective organisations before the outbreak of War and simply carried on their duties.

Whilst some of the medical Fellows found themselves in uniform and are mentioned in Chapter 4, others are mentioned here or in Chapter 9. Those mentioned here are Fellows who worked in establishments vital to the war effort, and those in Chapter 9 worked on projects vital to our overall well-being. Interestingly some Fellows were asked to re-evaluate sources of raw materials such as metals or fuel.

Those Fellows nearing retirement were often also in the Home Guard, some of which are listed, whilst others were indeed retired. Finally there were naturally some individuals who were too young for the War but went on to academic excellence in the post War years. Whilst this timeframe is outside the period of interest, some Fellows went on to the international scientific stage, perhaps with the award of the Nobel Prize, whilst others helped to re-build Britain. Some of these individuals will be mentioned for completeness.

There are a few exceptions to the general rules provided above. For example some Fellows were involved with developing armaments such as the Atom Bomb or the Bouncing Bomb. These Fellows, although either in industry or a Government Department, will be mentioned later Chapters.

Academic posts

None of our Universities escaped the effects of War. Some located in Central London or another major city were evacuated to less populated towns, others sadly bombed. All lost many students and some staff members to the fighting Services, or in the case of senior academics to 'think tanks' and Government Institutions. Three criteria were used to compile the information below:

- Those involved principally in research or administration.
- Those involved mainly in teaching.
- Those tasked with protecting data, samples or both.

Some of the Nation's senior academics offered their laboratories or their own time to help specific projects. Where appropriate some of these will also be mentioned.

Figure 5.1 shows details of some of our senior academics and their war-work.

Name	Occupation
Allen (Herbert Stanley)	Professor of Natural Philosophy and Director of Physics Research at the University of St. Andrews (1923–44).[1]
Astbury (William Thomas)	Director of Textile Physics Research Laboratory, and Reader, University of Leeds.[2]
Bawden (Sir Frederick Charles)	Head of Plant Pathology Rothamsted (1940), Deputy Director of Rothamsted (1950).[3]
Bennett (George Macdonald)	Appointed Chair of Organic Chemistry at King's College London. Moved the department to Bristol in 1939 and remained there until 1945.[4]
Berrill (Norman John)	Appointed Chair of the Department of Zoology at McGill University (1940).[5]
Bhabha (Homi Jehangir)	Reader (1940) and Professor (1942) at the Bangalore Institute of Science.[6]
Birch (Arthur John)	Remained at Oxford University during the War.[7]
Bradley (Albert James)	Assistant Director of Research in Crystallography, Cavendish Laboratory, Cambridge.[8]
Brown (William)	Professor of Plant Pathology, Imperial College of Science and Technology (1938–53), Head of the Botany Department from 1939.[9]
Browning (Carl Hamilton)	Professor of Bacteriology, University of Glasgow continued publishing throughout the War.[10]
Copp (Douglas Harold)	Worked at the Department of Biochemistry at Berkley (1939–50).[11]
Doodson (Arthur Thomas)	Associate Director of Liverpool Observatory and Tidal Institute.[12]

Name	Occupation
Erdös (Paul)	Theoretical mathematician who left Manchester for the Institute for Advanced Study in Princeton for the duration of the War.[13]
Falconer (Douglas Scott)	Moved from Queen Mary College to the Animal Genetics Section of the Animal Breeding and Genetics Research Organisation (ABGRO) in 1945.[14]
Fisher (Sir Ronald Aylmer)	Appointed to the Arthur Balfour Chair in Genetics at Cambridge (1943).[15]
Gates (Reginald Ruggles)	Research Fellow in Botany and Anthropology, Harvard University (1942).[16]
Goldsbrough (George Ridsdale)	Head of the Department of Mathematics (from 1942) of the University of Durham, and Chairman of the University's Joint Recruiting Board (1940).[17]
Gray (Joseph Alexander)	Chown Research Professor of Physics, Queen's University, Kingston Ontario during the War.[18]
Green (Albert Edward)	Lecturer in Mathematics, Durham Colls, University of Durham (1939–48).[19]
Holmes (Arthur)	Appointed Professor of Geology (1925), University of Durham and to the Regius Chair in Geology at Edinburgh (1943).[20]
Hörstadius (Sven Otto)	Appointed to the Chair in Zoology in Uppsala (1942).[21]
Hudson (Robert George Spencer)	Appointed Professor of Geology at the University of Leeds (1939). He resigned in 1940 spending the following two years as a Research Fellow of the University. He also assisted the Commissioner for Civil Defence in the North-western Region (1940–1942).[22]
Hughes (Edward David)	Chair of Chemistry at Bangor University (1943).[23]
Hutchinson (George Evelyn)	Appointed instructor in Zoology at Yale (1928) and Professor also at Yale (1945).[24]
Ingold (Sir Christopher Kelk)	Appointed Director of the Laboratories (1937) and head of the Chemistry Laboratory at University College (1939–44).[25]
Krishnan (Sir Kariamanikkam Srinivasa)	Mahendralal Sircar Research Professor of Physics at the Indian Association for the Cultivation of Science, Calcutta India.[26]
Lack (David Lambert)	Continued to publish throughout the War.[27]
Linderstrøm-Lang (Kaj Ulrik)	Appointed Professor and Head of the Chemical Department (1938).[28]
Maheshwari (Panchanan)	Married at 13 years old. Appointed Reader and Head of the new Department of Biology in Dacca where he remained until 1948.[29]
Marshall (Francis Hugh Adam)	Reader in Agricultural Physiology a position he held until his retirement in 1943.[30]

Name	Occupation
Mordell (Louis Joel)	Occupied the Fielden Chair of Pure Mathematics, University of Manchester (1923–1945).[31]
Offord (Albert Cyril)	Fellow, St John's College, Cambridge (1937–1940), Lecturer, University College of North Wales, Bangor (1940–1941), Lecturer, King's College, Newcastle-upon-Tyne (1941–1945).[32]
Pearsall (William Harold)	His six years in the Chair of Botany at Sheffield had been almost entirely wartime years.[33]
Pickard (Sir Robert Howson)	Became a member of the Senate of the University of London (1926) and remained a member of that body for twenty-two years.[34]
Pirie (Norman Wingate)	Published various papers during War.[35]
Ramachandran (Gopalasamudram Narayana)	Joined the Electrical Engineering Department of the Indian Institute of Science, Bangalore, in 1942.[36]
Shoenberg (David)	University Lecturer in Physics (1944–1952).[37]
Stephenson (Thomas Alan)	Appointed to the Chair of Zoology at Aberystwyth (1940).[38]
Struve (Otto)	Director Yerkes Observatory, University of Chicago and McDonald Observatory, University of Texas (1932–1947).[39]
Szwarc (Michael)	Engaged in research at the Hebrew University of Jerusalem and from which he gained a PhD in organic chemistry.[40]
Thoday (David)	Professor of Botany in the University College of North Wales, Bangor during the War.[41]
Timoshenko (Stephen Prokofievitch)	Remained at Stamford University throughout the War.[42]
Titchmarsh (Edward Charles)	Savilian Professor of Geometry at Oxford (1931–1960s).[43]
Tompkins (Frederick Clifford)	Lecturer then Senior Lecturer, Natal University, South Africa (1937–47).[44]
Trevelyan (George Macaulay)	Appointed the Master of Trinity College Cambridge (autumn 1940–June 1951).[45]
Turner (William Ernest Stephen)	Professor of Glass Technology, Department of Glass Technology at Sheffield University.
Tyndall (Arthur Mannering)	Henry Overton Wills Professor of Physics (1919–1948), Director of the H.H. Wills Physics Laboratory (1927–1940) and Acting Vice-Chancellor (1945) University of Bristol.[46]
Walshe (Sir Francis Martin Rouse)	Continued to publish throughout the War.[47]

Name	Occupation
Whittard (Walter Frederick)	Appointed Channing Wills Chair of Geology in Bristol (in 1937) remaining in that post for the duration of the War.[48]
Wood (Robert Williams)	Appointed Full Professor of Experimental Physics in 1901 at Johns Hopkins University and remained there for the rest of his life. Also consultant on the atom bomb and experimental work on cavity charges and their spectra.[49]

Figure 5.1 Some Fellows in Research or Administrative posts during the War.

In some cases the university departments were turned over to Government war-work. Under these circumstances a senior academic might have carried on at the university dealing with the normal administration whilst also playing host to scientists working for Government Departments. For example:

- Alexander Robertson FRS 'Professor of Chemistry, University of Liverpool. The university chemical laboratories were commandeered for various government agencies and a program was set up within the University, under Robertson's supervision, for work in connection with the Ministry of Supply.'[50]
- Sir James Irvine Orme Masson FRS 'Vice-Chancellor, University of Durham hosted a Ministry of Supply team working in the Chemistry Department. Dr Godfrey Rotter, then Superintendent of Research at Woolwich, negotiated an arrangement with Durham University in 1939 concerning the physical properties of amatols, and the correlation of the proportion, grain size, and grain shape of the ammonium nitrate with, ultimately, the explosive performance of the charge in large bombs and mines. A second line of enquiry dealt with fuse powders, and a satisfactory method was developed for the production of charcoal for gunpowder fuses, using available supplies of alder buckthorn in place of the wood which had hitherto been imported. He also acted as Chairman of the Sub-Committee on High Explosives of the Scientific Advisory Council.'[51]

Some Fellows, or scientists who went on to Fellowship, were just starting their careers during the War. At the time, these scientists were junior lecturers involved mainly with teaching, and included:

Name	Occupation
Besicovitch (Abram Samoilovitch)	Cayley Lecturer in Mathematics in the University of Cambridge (1927–1950).[52]
Bulman (Oliver Meredith Boone)	Lecturer in Palaeozology, University of Cambridge.[53]
Cartwright (Dame Mary Lucy)	Carried out a very full program of teaching and research, as well as a Commandant of the College Red Cross Detachment.[54]
Coulson (Charles Alfred)	Soon after the outbreak of War he appeared before a tribunal as a conscientious objector and was left to serve his country in his own way. He did this by doing an immense amount of lecturing (at Dundee University).[55]
Davenport (Harold)	Assistant Lecturer in Mathematics, Manchester University.[56]
Davidson (James Norman)	Appointed Lecturer in Biochemistry in the University College Dundee (summer of 1938) then Lecturer in Biochemistry in the University of Aberdeen (1940).[57]
Hutchinson (John)	Continued to publish papers on botany throughout the War whilst working at Kew Gardens.[58]
Keilin (David)	University Lecturer in Parasitology, Cambridge.[59]
Knowles (Sir Francis Gerald William)	Joined the staff of Marlborough College as Senior Biology Master (1938).[60]
Mitchell (George Francis)	Promoted to Lecturer in Geology (1940), elected to a Fellow of Trinity College (1944).[61]
Powell (Herbert Marcus)	University demonstrator and lecturer in chemical crystallography, Department of Mineralogy, University of Oxford (1934–1944).[62]
Rogosinski (Werner Wolfgang)	Appointed assistant (on £300pa) at Aberdeen University (1941), moved to Newcastle in 1945.[63]
Roughton (Francis John Worsley)	Lecturer in Physiology, Cambridge University (1927–1947).[64]
Schwarzschild (Martin)	Appointed lecturer at Columbia University Rutherford Observatory (1940–44) and then Assistant Professor (1944–1947).[65]
Stevens (Thomas Stevens)	Appointed Lecturer in Chemistry, Glasgow University (1925). Distinguished for his highly original researches in organic chemistry and as a teacher.[66]
Stewart (Sir Frederick Henry)	Appointed to a lectureship in geology in the Durham Colleges in Durham University (1943).[67]
Whittaker (Sir Edmund Taylor)	Professorship of Mathematics in the University of Edinburgh, taught for thirty four years until his retirement in 1946.[68]
Wilson (William)	Bedford College moved to Cambridge during the War where Wilson took part in the teaching of the Cavendish Laboratory.[69]

Figure 5.2 Some Fellows with heavy teaching commitments during the War.

For many people the threat of loss of personal possessions was a constant worry throughout the War. For some of the scientists with perhaps the only specimen known to exist it was more than a concern. In some cases rare objects were taken to secret locations in areas deemed less likely to receive a direct bomb hit. In other cases the scientists joined fire watch patrols. The following represent just some of these activities.

- Frederick Sydney Dainton FRS (Baron Dainton of Hallam Moors) 'was unfit for military service because of his short sight, in addition to a substantial teaching load in physical chemistry whilst at the same time as completing his PhD he was also responsible for organising air raid precautions and fire watching rotas.'[70]
- Professor James Peter Hill FRS 'had built up a very large collection of slides and materials during a lifetime of studying zoology and comparative anatomy. The slides on which he was actually working were taken to Hill's residence in Finchley, the best of his slides and the spirit material were taken along with specimens from the British Museum of Natural History to the cellars of the Rothschild Museum at Tring.'[71]
- Reginald Dawson Preston FRS 'spent much of his time on the protection of the university and on fire precautions across the whole of the northeast of England. As a result of this activity he did not get much continuous sleep. In addition to his fire duties he was a lecturer in the Botany Department (1936–1946).'[72]
- William Bertram Turrill FRS 'was placed in charge of a large portion of the collection from Kew Herbarium and Library which for safety had been moved to the basement of the New Bodleian Library in Oxford (1940).'[73]

Of course for some, the threat of fire and or bomb damage was only too real. The following is reproduced from the Royal Society Biographical Memoir for Alastair Graham FRS:

On the night of 25th September six incendiary bombs fell on Breams Buildings, and the Zoology Department was gutted. Graham was on fire-watch and was one of the first to pitch in to the shambles. Later he was seen with Dr Fretter salvaging animals' skeletons over a gaping hole in the floor where one wrong step would have carried them into an abyss. Weeks of salvage work followed and Royal Holloway became a sanctuary for the rescued material. By the end of October, and with the help of Dr Fretter (who was now officially at Royal Holloway full time), weekend replaced evening classes. Graham reciprocated by teaching at Holloway during the week. They ferried teaching material between the two colleges in rucksacks. In 1943 he became the Head of Department, and a year later was awarded a DSc (University of London). However, on 19 July Birkbeck was severely damaged by a flying bomb and the Zoology Department was gutted for a second time. Graham was on duty and, again risking life and limb, he clambered among the wreckage trying to stem a flood and extinguish a phosphorus fire.[74]

In extreme cases whole laboratories were totally destroyed through air attacks. For example Otto Renner FRS was Professor of Botany at Jena. His laboratory was destroyed in 1945.[75]

Of course many of the Nobel Laureates carried on their ground breaking work throughout all of the potential threats to their environment. Figure 5.3 lists just some of them. Not all Nobel Laureates will be mentioned here some were involved in the work to produce the atomic bomb, for example which will be mentioned in Chapter 7. Nobel Laureates mentioned here are not documented in detail elsewhere.

Name	Occupation	Nobel Prize
Born (Max)	Tait Professor of Natural Philosophy in the University of Edinburgh.[76]	Physics, 1954
Calvin (Melvin)	Assistant Professor at the University of California at Berkeley.[77]	Chemistry, 1961
Debye (Peter Joseph Wilhelm)	Appointed Professor and Chair of Department at Cornell University in 1940, retired in 1952.[78]	Chemistry, 1936
Eccles (Sir John Carew)	Became involved in a number of Committees and research projects dealing with the problems of vision, hearing, noise and communication in aircraft and tanks. He also actively participated with an army unit responsible for the supply of blood and serum for the armed forces.[79]	Physiology or Medicine, 1963
Hevesy (George de)	Worked at the Niels Bohr Institute in Copenhagen (from 1934). Associate of The Institute for Research in Organic Chemistry (from 1943).[80]	Chemistry, 1943
Ginzburg (Vitaly Lazarevich)	Academic life in the Lebedev Physical Institute in Moscow.[81]	Physics, 2003
Hahn (Otto)	Director of the Kaiser Wilhelm Institute (from 1928).[82]	Chemistry, 1944
Herzberg (Gerhard)	Research Professor, University of Saskatchewan, 1935–1945.[83]	Chemistry, 1971
Heyrovsky (Jaroslav)	Able to continue his research during the War and finished his large textbook on polarography.[84]	Chemistry, 1959
Hinshelwood (Cyril Norman)	Dr Lee's Professor of Inorganic and Physical Chemistry, University of Oxford (1937–1964), Chemical Defence Board of the Ministry of Supply (1940–1945).[85]	Chemistry, 1956

Name	Occupation	Nobel Prize
Houssay (Bernardo Alberto)	Professor of Physiology and Head of the Institute of Physiology in the University of Buenos Aires until the Government dismissed many academics in October 1943. Joined the Independent Institute of Biology and Experimental Medicine in 1944.[86]	Physiology or Medicine, 1947
Martin (Archer John Porter)	Worked at the Wool Industries Research Association in Leeds.[87]	Chemistry, 1952
Muller (Hermann Joseph)	Research Associateship (1941–1943), then Professorship (1943–1945) in the Department of Biology, Amherst College.[88]	Physiology or Medicine, 1946
Onsager (Lars)	Academic life at Yale.[89]	Chemistry, 1968
Pauli (Wolfgang Ernst)	Princeton University 1940–1945.[90]	Physics, 1945
Powell (Cecil Frank)	Appointed Lecturer in Physics in 1931, Reader in 1946 and Melville Wills Professor of Physics in 1948.[91]	Physics, 1950
Prelog (Vladimir)	Appointed Privatdozent (1942) and Professor (1945) at the ETH in Switzerland.[92]	Chemistry, 1975
Purcell (Edward Mills)	Joined the Radiation Laboratory at MIT and later led the Fundamental Developments Group.[93]	Physics, 1952
Raman (Sir Chandrasekhara Venkata)	Worked in the Indian Institute of Science throughout the War.[94]	Physics, 1930
Reichstein (Tadeus)	Professor in Pharmaceutical Chemistry, and Director of the Pharmaceutical Institute in the University of Basel (1938–1946).[95]	Physiology or Medicine, 1950
Schrödinger (Erwin)	Appointed the first Professor of Theoretical Physics at the newly formed Institute for Advanced Studied in Eire.[96]	Physics, 1933
Svedberg (Theodor)	Professor of Physical Chemistry in the University of Uppsala Sweden (1913–1949), instrumental in the creation of the Research Council for Technology (1942).[97]	Chemistry, 1926
Tiselius (Arne Wilhelm Kaurin)	Research Professor of Biochemistry, Uppsala Sweden (from 1938). Tiselius was asked by the Swedish Government to form part of the seven member Committee.[98]	Chemistry, 1948
Wieland (Heinrich)	Appointed Professor at the University of Munich in 1925 and remained there until the end of the War.[99]	Chemistry, 1927

Figure 5.3 Some Nobel Laureates and their academic War activities.

Civilian Consultants to the armed forces

Some Fellows acted as external consultants, called on when needed for specific knowledge or an activity needing resolution. For example:

- Alexander George Ogston FRS 'joined R.A. Peter's external research team of the Ministry of Supply. The team's aim was to find a non-toxic reagent to compete with tissue substances in converting mustard gas into non-toxic derivatives.'[100]
- John Lighton Synge FRS 'was appointed as a civilian consultant, with the official title Ballistic Mathematician, at Aberdeen (when not acting as a consultant he was Head of the Mathematics Department of the Ohio State University). It was in this capacity that he travelled to London early in 1944 attached as scientific assistant to Colonel Schwarz, the Armament Officer for the US Army Air Force in Europe.'[101]
- Nevil Vincent Sidgwick FRS 'was an unpaid consultant to the Department of Explosive Supplies.'[102]
- Sir Alan Walsh FRS 'determined which metals were being used in enemy bombers that had been shot down, from which War economists worked out how the German war effort was progressing.'[103]
- Sir Thomas Henry Havelock FRS 'was a mathematician and Sub-Rector of King's College the Newcastle Division of Durham University (now Newcastle University). One of Havelock's interests was his work on ship resistance. Accordingly, he had close contact with the Department of Naval Architecture.'[104]
- Solomon Lefschetz ForMemRS 'served as a consultant to the US Department of the Navy at the David Taylor Model Basin.'[105]

Industry/Government Establishment based Fellows

There were many Fellows who worked in either an industrial company or in a Government Establishment as a paid employee. In all cases the Fellows were working in a reserved occupation. Where the Fellow worked abroad, which will be documented in a separate table, the equivalent terminology in that country for our reserved occupation is assumed.

Name	Occupation
Akers (Sir Wallace Alan)	Technical manager in the recently formed Imperial Chemical Industries (from 1928) joining the Board of the company in January 1941. At that time the Lord President of the Council established a special organisation dealing with the whole problem of atomic energy and its various aspects, including R&D. Akers became the first Director his services being lent by his company.[106]
Chesters (John Hugh)	Head of Ceramics Section, Central Laboratories United Steel Co (1934–1945).[107]
Crombie (Leslie)	Worked as a laboratory assistant in the paints laboratory at the Admiralty Chemical Department in Portsmouth dockyard.[108]
Desch (Cecil Henry)	Joined W. Cubbit and Co Grays Inn Road with whom he remained for 52 years, attaining the position of Chief Surveyor and for the last few years was a Director in the Company.[109]
Dorey (Stanley Fabes)	Joined Lloyd's Register of Shipping as a Ship and Engineer Surveyor (1919), appointed Chief Engineer Surveyor (1932).[110]
Eckersley (Thomas Lydwell)	Worked in the Marconi's Wireless Telegraph Co. Ltd. (1918–1946).[111]
Fleck (Alexander: Baron Fleck of Saltcoats)	Appointed Chairman of ICI's Billingham Division in 1937. During the War, and at Government expense, a second plant was built on Heysham to treat aromatic petroleum gas oil. Special aviation spirit additives and components were also produced at Billingham.[112]
Frankel (Sir Otto Herzberg)	Appointed Chief Operating Officer of the Wheat Research Institute in 1942.[113]
Gee (Geoffrey)	Worked at the British Rubber Producers Research Association (1938–53).[114]
Harris (John Edward)	Worked on the Corrosion Sub-Committee of the Iron and Steel Institute investigating settlement of marine organisms on the hulls of ocean-going ships (barnacles etc.). Appointed Professor of Zoology University of Bristol in 1944.[115]
Hume-Rothery (William)	Warren Research Fellow of the Royal Society worked on the modern theory of alloys.[116]
Pfeil (Leonard Bessemer)	Joined The Mond Nickel Company as assistant manager of the Research and Development Department (1930). The major wartime contribution of the laboratory was the development of heat resisting alloys for use in gas turbine engines.[117]
Simpson (Sir George Clarke)	Came out of retirement and took charge of Kew Observatory and administratively of Eskdalemuir , Aberdeen and Lerwick Observatories and of the Edinburgh office. He was also a member of the Aeronautical Research Committee (Air Ministry) and of the Radio Research Board (DSIR, 1924–1946).[118]

Name	Occupation
Stoll (Arthur)	Appointed Director (1933) and elected Vice-President (1935) of the Sandoz Company.[119]
Stopford (John Sebastian Bach, Baron Stopford of Fallowfield)	Upon his retirement he was appointed Vice Chairman of the Nuffield Foundation (1943).[120]
Uvarov (Sir Boris Petrovich)	Anti-locus work latterly Director of the Anti-Locus Research Centre of the British Museum.[121]
Wynne (Charles Gorrie)	Designed lenses initially for TT&H Company until 1943, then the Wray (Optical Works) Company. The Wray company had taken up the manufacture, under Government contract, of telephoto lenses with a focal length of 36 inches (1942). These lenses enabled the RAF photographic reconnaissance units and the photo-interpreters to make the crucial high-altitude photographic detection of the development of the V weapons and the jet and rocket fighters that Germany developed in the later stages of World War Two.[122]
Young (Alec David)	Divided his time between the Seaplane Tank Laboratory, the Wind Tunnels Section and the Flight Section of the Aerodynamics Department of the Royal Aircraft Establishment (RAE), Farnborough (from the summer of 1936).[123]

Figure 5.4 Some Fellows working in England.

These are but a few of the many Fellows who worked either in industry or for a Government Department. Whilst all of these Fellows had interesting occupations, perhaps that of Sir John Henry Gaddum FRS, (see figure 5.5), was particularly interesting. When War broke out, Gaddum was working at the Chemical Defence Research Station, Porton with Sir Joseph Barcroft FRS, C.G. Douglas FRS, Sir Charles Arthur Lovatt Evans FRS and Sir Gordon Roy Cameron FRS. He was partly responsible for deciding on the substance which should be carried by British agents to be used for rapid self-destruction in a serious emergency, sometimes known as the suicide pill. He later accepted the Chair of Materia Medica in the University of Edinburgh (1942).[124]

Figure 5.5 John Henry Gaddum. (©
Godfrey Argent Studio. Reproduced with
permission)

Some Fellows worked abroad during the War years.

Name	Occupation
Axelrod (Julius)	Chemist, Laboratory of Industrial Hygiene, New York City Public Health Department (1935–1946).[125]
Bhatnagar (Sir Shanti Swarup)	Appointed the first Director of the Board of Scientific and Industrial Research established by the Indian Government.[126]
Bowden (Frank Philip)	Appointed Officer-in-Charge of a section of the Council for Scientific and Industrial Research (forerunner of the Commonwealth Scientific and Industrial Research Organisation in Australia) from 1 November 1939.[127]
Casimir (Hendrik Brugt Gerhard)	Professor of Physics, Leyden (1939–1942), Philips Company Research Laboratory, Eindhoven (1942 onwards).[128]
Ewing (William Maurice)	Research Associate (1940–1944) for the Woods Hole Oceanographic Institution.[129]
Gregory (Frederick Gugenheim)	Advised on physiological problems of cotton cultivation under irrigation in the Sudan (1928–1946) at the request of the Director of Agriculture, Gezira Research Station.[130]

Figure 5.6 Fellows working abroad during the War.

Fellows in medical posts

Many Fellows were trained in the medical or veterinary professions or related occupations. Some Fellows occupied positions in institutes working long hours in their chosen field. Others worked on solving specific wartime problems. Those mentioned here are the former category. Fellows who became involved in specific wartime projects will be mentioned in Chapter 9.

Name	Post
Bancroft (Henry)	Professor at the Royal Veterinary College in London (1944).[131]
Bartlett (Sir Frederick Charles)	Psychologist who served on the RAF's Flying Personnel Research Committee, the Medical Research Council and in many other positions. The MRC founded an Applied Psychology Research Unit under Craik's direction in 1944 (he was a member of Bartlett's department).[132]
Elton (Charles Sutherland)	Increased the Bureau of Animal Population's staff to about 15 people and undertook research on rodent control.[133]
Gale (Ernest Frederick)	Research Fellowship at St John's College (1941–1944), appointed to the Scientific Staff of the MRC (1943).[134]
Huggett (Arthur St. George Joseph McCarthy)	Appointed to the Chair of Physiology at St Mary's Hospital Medical School (in 1935 remaining until 1964).[135]
Loewi (Otto)	Then Research Professor of Pharmacology, Medical School of the New York University (from 1940).[136]
Mandelstam (Joel)	Worked as a research assistant to Dr Joseph Gilman at the Medical School in Johannesburg (from 1942).[137]
Sir Wilfrid Edward le Gros Clark	Professor of Anatomy in the University of London at St Thomas's Hospital.[138]
Smith (Wilson)	Professor of Bacteriology at the University of Sheffield (1939–1946).[139]
Wilkie (Douglas Robert)	Entered University College London (UCL) to study medicine (1940). Won the Rockefeller Scholarship to Yale University, New Haven, Connecticut, for the last year of his medical education. The Physiology Department of UCL appointed him to an assistant lectureship in 1945 when he was 23 years old.[140]

Figure 5.7 Some Medical Fellows.

Sadly, some Fellows met with accidents or were medically unfit for active duty. Their respective Royal Society Biographical Memoirs comment:

a) Alfred Gordon Gaydon FRS. 'During a chemistry experiment an explosion occurred exposing him to a hail of glass shards. His right eye was so severely damaged and thought to be infected and was removed some weeks later, and the crystalline lens of the left eye was punctured by small pieces of glass and formed a cataract. Gaydon was totally blind for some six months. During the early 1940s Gaydon collaborated with George Wald, the Nobel Prize-winning biologist, who discovered the presence of vitamin A in the retina and went on to show that the three retinal pigments involved in human colour vision are all combinations of vitamin A with one of three different proteins. After Gaydon's observations, which as a spectroscopist he was uniquely placed to interpret, Wald performed important work with cataract patients on the sensitivity of the human eye to the wavelengths of the spectrum.'[141]

b) Thomas Stanley Westoll FRS. 'At the outbreak of War in 1939 he tried to enlist in the army. He passed the medical and was drawing his uniform when he asked for a special boot which drew attention to his disability.'[142]

c) William Ogilvy Kermack FRS 'appointed in charge of the Chemical Section of the Royal College of Physicians Laboratory in Edinburgh (1921). On the evening of Monday 2nd June 1924 Kermack was working alone in his laboratory when the preparation he was working on exploded driving caustic alkali into his eyes. After two months in hospital he was discharged permanently blind.'[143]

The search for raw materials

Some Fellows were involved in geological surveys looking for valuable raw materials:

Resource	Name	Activity
Coal	Richey (James Ernest)	Re-assessment of the potentialities of the Canonbie coalfield, and a structural study of the oil shale field near West Calder.[144]
Coal	Stubblefield (Sir (Cyril) James)	The need for more accurate knowledge of Coal Measure fossils to improve classification and correlation of the strata was urgent, and Stubblefield had to divert his attention from the Lower Palaeozoic strata to these younger rocks.[145]

Resource	Name	Activity
Magnesium	Fearnsides (William George)	Located hard dolomite of suitable purity in sufficient abundance and sufficiently close to an acceptable supply of sea water. A new processing plant solely for magnesium metal purification was erected on the Cumberland coast.[146]
Oil	Lees (George Martin)	Searched for oil in Britain and its discovery in sufficient quantity to pay for the costs of exploration.[147]
Sand and gravel	Hawkins (Herbert Leader)	Appointed Geological Adviser to the Civil Defence Area VI in 1939. Initiated a rapid survey of sand and gravel resources, for sandbags etc.[148]

Figure 5.8 Some activities of a few Professors of Geology.

Magnesium was used in incendiary bombs and light alloy castings for example.

At least one of the Fellows was called in as a consultant on the collapse of the Chingford Dam in the Lea valley, Essex. The Royal Society Biographical Memoir for Sir Alec Westley Skempton FRS contains the following:

> Skempton had been at the Building Research Station for only a few months when the Chingford Dam in the Lea valley, Essex, which was then under construction, collapsed. Similar dams in the Lea valley had previously been constructed quite safely, so the reason for the collapse was not immediately apparent. The BRS was asked to advise, and the soil mechanics group, of which the young Skempton was rapidly becoming a leading member, quickly came to the conclusion that the problem was the incomplete consolidation of the alluvial clay foundations. This elegantly explained the stability of the earlier dams, which had been constructed more slowly in the horse-and-cart era (thus allowing time for consolidation) than was the case with the mechanised plant used at Chingford. Terzaghi was brought in by the contractor to confirm the diagnosis, leading to a long association between Terzaghi and the BRS group.[149]

Civil Defence

Civil Defence, in its broadest meaning, was a part of everyday life. Every person in the country, of any age, was able to pass on to the authorities every incident deemed of importance. Some of these activities deserve special mention in that they took over the majority of a person's working day, or they were of particular significance. For example who can forget the newspaper pictures of the fire watchers on the roof of St Paul's Cathedral in London during the height of the blitz?

In many cases the individual's stories are less dramatic, though equally important – for the country had a lot to lose should any building be destroyed. Of course fire watching was only one part of the overall strategy. The Home Guard was just a vital to the countries wellbeing as was fire watching duties. Home Guard Fellows included:

- Colin Eaborn FRS 'signed on for the Local Defence Volunteer Force (later the Home Guard in April 1940.'[150]
- Francis William Rogers Brambell FRS. 'Whilst Dean of the Faculty of Science (1939 to 1943), Brambell was also a founder member of the Home Guard (since its inception in 1940), continuing until its disbandment. He was commissioned 2nd Lieutenant in 1941.'[151]
- William Cochran FRS 'applied for a commission in the Technical Branch of the Royal Air Force Volunteer Reserve under the Hankey Radio Training Scheme. Unfortunately due to some administrative failure, the Ministry of Labour and National Service, thought Cochran had withdrawn his candidature. He hadn't. He did not then, nor since, understand the basis of the letter from the Ministry of Labour and National Service, so his War service was confined to the Home Guard, of which he was a member for the whole of its existence.'[152]

Fire watching Fellows included:

- Edward Battersby Bailey FRS 'took part in fire watching and V1 warning duties whilst Lieutenant commanding the Geological Survey and London Regional Unit of the 58th County of London Battalion (until 1942).'[153]
- George Dixon Rochester FRS. 'Initially deployed as a radar scanner to Saxton Wold, near Scarborough, Rochester took over the duties of University Fire Officer, responsibilities for training Army and Royal Air Force cadets and Civil Defence Adviser for the northwest and later for the northeast of England.'[154]
- Herbert Harold Read FRS 'was heavily involved in Civil Defence and in fire watching duties at Imperial College in addition to his appointment to the Chair of Geology at Imperial College (1939).'[155]

Too young and too old

Even though some of the Fellows were in reserved occupations there were some who retired during the War. That is not to say that they did not then take up a War related activity, merely that they retired from their full time paid work. For example:

- Herbert John Fleure FRS 'Professor of Geography and Anthropology in Victoria University, Manchester – retired in 1944.'[156]
- Vernon Herbert Blackman FRS 'retired as Director of the Research Institute of Plant Physiology at Imperial College in 1943.'[157]
- William Kingdon Spencer FRS 'retired from the post of Inspector of Schools in 1938. At one point he was appointed HMI for Ipswich, East Suffolk and Lowestoft and also Chief Inspector for England and Wales.'[158]

All Fellows too young to serve in the War or at home helped to shape the post War world in which we now live. As a practicing industrial scientist for many years there are, at least for me, some stand-out names of scientists who lived through the War, shaping university teaching or industrial research from their research groups perhaps thousands of miles away from where I lived. For example:

Born on 9 April 1930, Frank Albert Cotton ForMemRS went on to write a book with Sir Geoffrey Wilkinson FRS (incidentally the Chemistry Nobel Laureate for 1973) which became a standard undergraduate textbook – at least in the 1970s.

Sir John Anthony Pople FRS (Chemistry Nobel Laureate 1998) co-wrote a computer program called the PPP, one of the early forays into molecular orbital calculations I used in industry in the 1980s.

Martyn Christian Raymond Symons FRS worked for many years at Leicester University. I remember Martyn from my time visiting him in 1981 during a job interview.

Who can forget William David Ollis FRS, a memorable addition to any conference in the 1970s and 80s.

Professor Charles Wayne Rees FRS retired from Imperial College. One of the finest organic chemists I met regularly over a thirty year period from my doctoral years at Salford University until his untimely death on 21 September 2006.

Additionally there are many living Fellows such as my own post-doctoral supervisor Professor Alan Roy Katritzky FRS who has shaped the world of organic chemistry and helped to develop the talents of almost 850 students (included doctoral and post-doctoral workers).

Each scientist will have his or her list of personal stories from those Fellows too young for active service during the War. These are just some of mine. In all cases these Fellows shaped our present and will our future.

6   Allied concealment, enemy detection

Allied Concealment

William Edward Curtis FRS was an interesting Fellow. The following appears in his Royal Society Biographical Memoir and/or *Who was Who* entries:

> Professor of Physics at King's College Newcastle upon Tyne (1926–55). Appointed a Scientific Advisor to the Ministry of Home Security where he was put in charge of the Civil Defence Camouflage Establishment at Leamington with the title of Director of Camouflage and Decoy. This establishment was responsible for the camouflage of factories and civil airfields and was originally staffed by artists. Work was first concerned with daytime camouflage and later with night camouflage. One of the most successful pieces of work was creating a decoy of Sheffield which diverted numerous bombing raids from that city. As the intensity of the air attacks diminished the work of the establishment was so reduced that in 1942 Curtis was transferred to Princes Risborough as one of the Scientific Advisers to Reginald Stradling in the Research and Experiments Branch of the Ministry of Home Security (1942–43). The other advisers at this time were Lord Zuckerman FRS and Professor J. Bernal FRS, the group being known in certain quarters as 'The Three Wise Men'. Their work was concerned with the interpretation of aerial photographs of bomb damage to targets in enemy and enemy held territory with a view to assessing the damage to the German War potential, from a careful analysis a large body of information, collected by the Ministry of Home Security, on the effects of German bombing on certain targets in this country. In 1943 Curtis was transferred to ARDE at Fort Halstead where he became Superintendent of Applied Explosives, a post he held to the end of the War (1943–45).[1]

Some of the Memoranda of the Camouflage Committee were published as a series of standalone documents. For example they included:

1) Camouflage of oil installations.
2) Camouflage of roads.
3) Camouflage treatment of aerodrome runway surfaces.
4) Concealment of water.
5) Texturing of roofs.
6) Treatment of sacred ground.
7) Treatment of scrim for fire prevention.

By way of an example, some extracts from the *Concealment of water* appears below.

Water Camouflage: Technical Methods

To conceal water it is necessary to prevent it reflecting the sky, the sun or the moon. There are only two methods of doing this. The first is to remove the water; the second is to cover it.

1) Drainage – this, of course, is the best solution. Even if a pond cannot be completely drained, its area can often be considerably reduced. This reduced area of water may not greatly resemble the original pond and, if it should be decided that even the reduced area must be hidden, it presents a smaller problem.
2) Covering of water – has been the subject of many experiments in this country though only one large scale job had been completed. Most of the experiments have been with a thin film of coal dust and oil or with fabrics such as hop lewing (coir), onion bagging (jute) and a special camouflage material supported above the water in various ways.
3) Capacity of covering material.
4) Types of covering material.
5) Types of supporting structure.
6) Types of floats.
7) Non-floating structures.
8) Rafts.
9) Low rafts (experimental types).
10) Canopy of floating supports.
11) Concealment of small craft.
12) Snow and ice.
13) Report on three experiments with rafts.
14) Enemy practice – in enemy countries, water concealment has been carried out on an enormous scale. The best known instance is that of the Binnen Alster in Hamburg but elsewhere even larger areas have been covered. Details of methods are, of course lacking, but some lakes and canals have been rafted, and others covered with canopies of

opaque material supported on posts in the water or slung between the banks.

15) Coal dust and oil – experiments have been carried out with two other methods of concealing water. The first is the covering of the surface with a thin film of coal dust and oil. This was found very successful in preventing any reflection of the sky, even moderate winds, however, blew the film to one side of the pond or canal and no solution of this difficulty was ever discovered. Screening, of course, can be erected to keep the wind away from the water surface, for such screening to be effective it must be almost as elaborate as the structure needed for rafts.

16) Water plants – the second method is the encouragement of water plants. Such plants as duckweed and azolla are not rooted in the soil but float in the water, covering sometimes very large areas. The concealment given is, however, illusory. From the ground it may appear excellent, but, from the air, water covered with such plants reflects the light strongly.[2]

Most of the memoranda mentioned above were published in 1941. The following year the Department published a Policy for Concealment. Some opening general remarks from which are:

The appreciation for the concealment required for any particular target must be based on a number of factors which vary in each case. These include in particular the importance and nature of the target, its position, the form of attack likely to be developed, the nature of existing defences of all kinds, meteorological factors etc. It is recognised that air observation must be limited, partly because of the shortage of aircraft available for this work, but chiefly because weather conditions, especially at night, render opportunities for observations few and far between. Additional difficulties in observation arise owing to the following variable factors:

- Conditions of visibility which are seldom the same upon any two given occasions.
- The temperament, experience and acuteness of night vision of individual observers.
- The differences in aircraft likely to be used for reconnaissance.

A combination of weather conditions, limited aircraft and other factors consequently make it advisable, if concealment is to be carried out within reasonable time, first for the Air Staff to amplify their general policy instruction, and secondly for directing staffs responsible for concealment to adopt methods which, though varying for each individual target, aim at minimising conspicuousness generally. Air observation should be used, as and when available, and may be essential in certain cases, but a great deal of work can and must necessarily be done without it.[3]

Further in the report there is a list of classifications of concealment viz:

- Class A – maximum night and day concealment by full netting of a few super-important targets, small enough to be dealt with in this manner.
- Class B – maximum or considerable night and day concealment by the use of some netting and careful camouflage designed to break abrupt outlines, for a few super-important targets too large to be fully netted.
- Class C – good concealment by night and a lesser standard of concealment by day for such targets whose location and importance justifies this extra standard.
- Class D – normal concealment by night.

  Note: in many cases concealment will be unnecessary

Bletchley Park/Intelligence work

Unlike 'Room 40', its World War One counterpart, Bletchley Park is more widely associated with code breaking – at least it was after the end of the War. Perhaps inevitably, both organisations were strongly influenced by Fellows, Bletchley Park more so given the number of them and the post War media exposure afforded this otherwise unremarkable country house in Buckinghamshire.

A City of London financier, Herbert Samuel Leon, bought over 300 acres of land in 1883 because the land was accessible by the London and North-Western Railway line that passed through Bletchley. He developed sixty acres into his country estate and built Bletchley. Captain Hubert Faulkner, bought the site following the deaths of Sir Herbert and Lady Fanny Leon intending to develop the whole site as the mansion had fallen into a state of disrepair.

The Government bought Bletchley Park in 1938 before it was developed as a base for the Government Code and Cypher School (G.C. and C.S.), then based in London. The rationale being it was safe from the threat of bombing raids on London, it had major rail and road links and it had teleprinter connections to all parts of the country.

Initially commanded by Alastair Denniston, the Park was given the cover name Station X, being the tenth of a large number of sites acquired by MI6 for its wartime operations.[4]

Using the roll of honour available on the Bletchley Park website,[5] and the Fellows database it is possible to determine which Fellows worked at Bletchley Park, and which hut was their predominant location, see figure 6.2. Interestingly, Turing's entry in the roll of honour is somewhat sparse, hence the website reference.

Figure 6.1 Bletchley Park. (Author's photograph)

Name	Activity
Jones (Reginald Victor)	Air Ministry Civilian. Bletchley Park October–November 1939. Hut 3. Detached from MI6 Air Section to acquaint himself with Bletchley Park's capabilities and senior personnel. Later Scientific Adviser to Air Ministry and Director of Scientific Intelligence.
Newman (Maxwell Herman Alexander)	FO Civilian, TSAO. Bletchley Park September 1942–May 1945. Research Section. Block F. Produced mechanical methods to break German enciphered printer. Set up Newmanry to exploit Robinson and Colossus.
Rothschild (Dame Miriam Louisa)	FO Civilian. Bletchley Park. Possibly Hut 6.
Rees (David)	Civilian mathematician working in Hut 6.
Turing (Alan)	Hut 8 which deciphered Naval and in particular U-boat messages, became a key unit at Bletchley Park.[6]
Tutte (William Thomas)	FO Civilian, TSAO. Bletchley Park May 1941–1945. Mansion and Hut 5. Research Section. Worked out internal structure of German Lorenz enciphered teleprinter ('Tunny') and devised technique for establishing wheel patterns used.

Figure 6.2 Fellows working in Bletchley Park.

Arguably one word symbolises the wartime activities at Bletchley Park more than any other, that of Enigma – a German encryption device, see figure 6.3. Yet it is doubtful that an Enigma machine was ever installed in the main house during the War because the grounds were used to construct huts, where most of the work took place.

Although some of the huts have been restored some have remained closed to the public pending opportunity and funds for further restoration.

The hut system of working had distinct advantages in that access to the huts could be restricted. Furthermore, working on the principle of 'need to know', information concerning the activities of any one hut was also restricted, further preventing loose talk of activities.

Figure 6.3 An Enigma machine. (Author's photograph)

The Enigma machine worked as follows:

1) Each wheel has on one side a single set of 26 pins and on the other side 26 flat contacts, these two contacts are wired to each other. The reflector has 26 pin contacts wired to each other.

2) When a key is pressed the current enters the machine at a current entry plate (ie one of the 26 contacts on the current entry disc) goes through the three wheels to the reflector, then comes back through the three wheels to the current entry plate and thence to the lampboard on which a light then shows up.

3) The stecker plugs connect the keyboard and lampboard to the current entry plate, the connections being reciprocal ie if A and G are connected by a strecker plug then the connections AG and GA are made both between keyboard and the entry disc and also between current entry disc and lampboard.

4) When a key is pressed the turnover pawl engages in the rim of the wheel next to the current entry plate and turns the wheel over one place, so that if the same letter were again encoded the path travelled by the current through the machine would be a different one.[7]

Figure 6.4 is a sketch of the above description.

Prior to the outbreak of the War, the Poles developed a machine they called a Bombe, intending to use it to crack Enigma. Unfortunately, the Bombe as described by the Poles was not up to the task.

Arguably, the most well-known of the 'Bletchley Fellows' is Alan Turing FRS, see figure 6.5. The Enigma machine will forever be associated with Turing, yet he didn't actually design an Enigma machine. Born on 23 June 1912, Alan Mathison Turing FRS was a Cambridge mathematician before the War.[8]

In his *Cryptographic History of the work on the German Naval Enigma*, C.H. O'D Alexander commented of Turing's input:

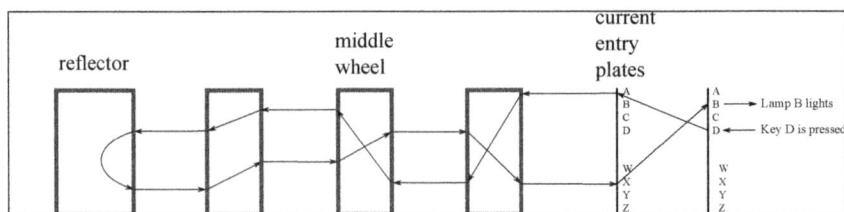

Figure 6.4 Schematic diagram of the working of an Enigma machine. (National Archives HW 25/1)

When the War started probably only two people thought that the Naval Enigma could be broken – Birch, the Head of the German Naval Section and Turing, one of the leading Cambridge mathematicians who joined G.C. and C.S. for the duration of the War. Birch thought it could be broken because it had to be broken and Turing thought it could be broken because it would be so interesting to break it. Whether or not these reasons were logically satisfactory they imbued those who held them with a determination that the problem should be solved and it is to the pertinacity and force that, in utterly different ways, both of them showed that success was ultimately due. Turing first got interested in the problem for the quite typical reason that 'no one else was doing anything about it and I could have it to myself'. He started where the Poles left off and set to work to discover how the indicating system worked using the information provided by the 100 or so messages in the period 1st–8th May1937, whose starting positions were known.[9]

Figure 6.5 Alan Turing (© Godfrey Argent Studio. Reproduced with permission)

Working with another Cambridge mathematician, W.G. Welchman, Turing and Welchman realised that they needed to improve or radically change the Bombe. Most commentators record the valuable contribution made by Welchman, but the input from Turing is universally regarded as the critical factor in the Bombe's success. The Bombe was used to predict which combinations could be ruled out thereby reducing the workload considerably. As far as is known all of the wartime Bombes were destroyed after the War. The British Tabulating Machine Company at Letchworth built over 200 Bombes![10]

Figure 6.6 shows the timeline of Bombe introduction, demonstrating the pace of change and highlighting the sense of urgency in the decryption effort.

Date	Bombe number	Number of wheels
18 March 1940	1	3
8 August 1940	2	3
25 Jan 1941	3	3
18 March 1941	4	3
16 May 1941	5	3
9 June 1941	6	3
15 August 1941	7	3
11 Sept 1941	8	3
9 Oct 1941	9	3
22 Oct 1941	10	3
23 Oct 1941	11	3
7 Nov 1941	12	3
26 Nov 1941	13	3
8 Dec 1941	14	3
20 Dec 1941	15	3
24 Dec 1941	16	3

Figure 6.6 Dates for the first 16 bombe introductions. (National Archives HW 25/31)

By 22 September 1944 the 141st Bombe was brought into service. Four wheel Bombes were developed in parallel to the introduction of the three wheel versions, the first four wheel Bombe being introduced on 22 March 1943. The rate of introduction of the four wheel Bombe was, if anything, at a greater pace than that of the three wheel versions. By 2 May 1944 the 37th four wheel Bombe was introduced which was the 6th W.W. type. So not only were the number of wheels increased from three to four, there were also variations/modifications being introduced.

Figure 6.7 An internal view of the replica Bombe in Bletchley Park. (Author's photograph)

The basic theory was that common words such as proper nouns, or weather reports of short sentences, were relatively easy to guess. Additionally a feature of Enigma machines was that no letter could be reproduced as itself. These guesses were called cribs. The cribs determined the wiring round the back of the machine enabling the Bombe to rule out incorrect Enigma settings. Cumbersome though this was, the system worked.

Figure 6.8 shows a schematic diagram of the work involved. The better the initial crib, the smaller the number of cycles needed to determine the possible Enigma settings, all of which took place in the machine room.

By mid-1941 Turing's section became a key unit at Bletchley Park. Hut 8, which deciphered Naval and in particular U-boat messages, was the nerve centre, until 1 February 1942, when the Atlantic U-boat Enigma machine was given an extra complication. It was to take just over a year before the logical weaknesses in the changed U-boat system were detected, restoring U-boat Enigma decryption from then until the end of the War.

Even though Turing left Hut 8 during the War, the number of messages Hut 8 personnel decrypted continued to rise. By way of example, figure 6.9 shows the daily registered traffic for the week ended 5 November 1943, where Dolphin, Shark, Turtle, Porpoise and Grampus are code names for various German sources.

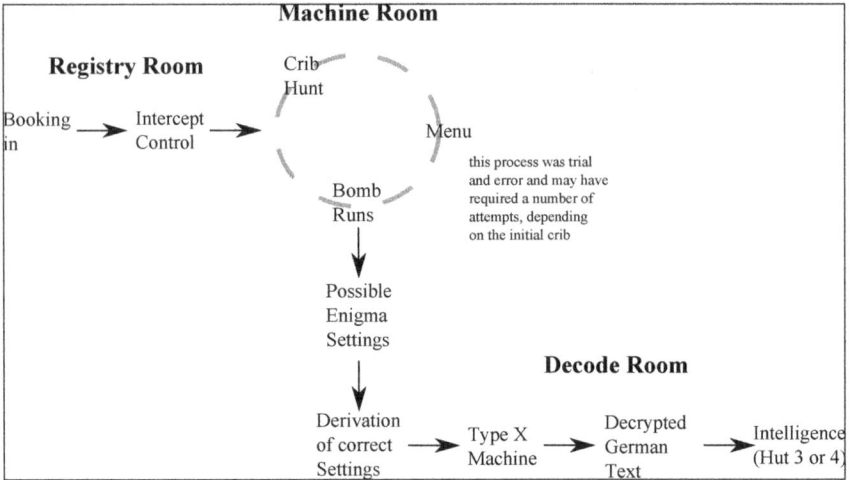

Figure 6.8 The process of decryption for Enigma messages. (Author's sketch)

	Dolphin	Shark	Turtle	Porpoise	Grampus	Total
29/10/43	576	77	40	385	130	1208
30/10/43	489	65	40	347	151	1092
31/10/43	483	82	35	380	231	1211
1/11/43	579	79	50	311	192	1211
2/11/43	629	53	38	316	200	1236
3/11/43	637	47	22	347	261	1314
4/11/43	596	62	26	357	220	1261
Daily average 29/10–4/11	570	66	36	349	198	1219
Daily average for October 1943	582	69	60	322	130	1163
Daily average for November 1942	423	123	–	275	–	821

Figure 6.9 Daily registered traffic. (National Archives HW 77/7)

From late 1940 onwards, the Turing–Welchman Bombe made reading of Luftwaffe signals routine.[11] Although routine, there were still three processes needed before a message arriving from Station Y could be sent out of Hut 3 or 4 as fully decoded messages. The need for secrecy and 24/7 cipher acquisition necessitated a series of shifts known as 'Watches'. There was also a need for an infrastructure dealing with the organisation of the cipher policy which was a Cabinet Committee. Appendix VII details the Committee structure, and its members at the end of the War.

Whilst Turing was busy with Enigma work, John Tiltman was equally busy with ciphers from the Lorenz machine. Rejected by Turing during the interview process for a message-code breaking team, William Thomas Tutte FRS was recruited by Tiltman for the Lorenz research section (in May 1941).

Made by the Lorenz Company, and even more complex than Enigma, the Lorenz machine was used exclusively for communications between German Army Field Marshals and their Central High Command in Berlin. Unlike Enigma, the Lorenz machines were not portable. The code breakers called the coded messages 'Fish' and the machine 'Tunny'.

The Bletchley Park website describes the Lorenz thus:

> The Lorenz used the 'International Teleprinter Code', in which each letter of the alphabet is represented by a series of five electrical impulses. Messages were enciphered by adding, character by character, a series of apparently randomly generated letters to the original text. Crucially, to decrypt the enciphered message, the receiving Lorenz simply added exactly the same obscuring letters back to the cipher text. The obscuring letters were generated by Lorenz's 12 rotors, five of which followed a regular pattern, while another five followed a pattern dictated by two pin wheels. Cracking Fish again relied on determining the starting position of the Lorenz achine's rotors.[12]

The first Fish messages were decoded by John Tiltman 1941. Tiltman used hand-methods that relied on statistical analysis. However, Lorenz machines were upgraded introducing complications which made it virtually impossible to break Tunny by hand. Clearly a machine, was needed to break Tunny, a task assigned to Dr Max Newman and his team. Maxwell Herman Alexander Newman FRS was the Fielden Professor of Mathematics, Manchester University.[13] He assembled a team which produced the first machine at the Post Office research department at Dollis Hill, nicknamed 'Heath Robinson'.

'Heath Robinson' was built by Tommy Flowers, a Post Office Electronics Engineer. Flowers went on to build Colossus, a replica of which is shown as figure 6.10, a much more reliable and faster machine using 1,500 thermionic valves, the first of which was delivered to Bletchley Park in December 1943.

Perhaps it is worth recording here some of the engineers and technicians who worked on building the first Colossus. These names appear in an appendix to a document titled *A Technical Description of Colossus I* written by D.C. Horwood.[14]

Figure 6.10 A replica Colossus. (Author's photograph)

- Mr T.H. Flowers is credited with the idea and the design for the original Colossus and the versions which followed closely after it. His official title was Staff Engineer in charge of the Signalling Laboratory.
- Mr S. Broadhurst worked directly under Mr Flowers at Dollis Hill and from the initiation of the Colossus project was almost entirely responsible for the design of the electromechanical element of the machine.
- Mr Lynch was a senior engineer at Dollis Hill in 1943 concerned with the design of the tape reader assembly of Colossus which was 'sub-contracted' within Dollis Hill to another group.
- Dr A.W.M. Coombes joined the Colossus project about October 1943 as an Executive Engineer under Flowers. He assisted in the design and super-vised the production of the second and subsequent models of Colossus.
- Mr W. Chandler joined the Colossus project at Dollis Hill in early 1943 and was involved under Flowers in the design and construction of all the Colossus machines including Colossus I.

Messers T. Hauser, F. Fensom, B. Clayton, Tompson, J. Cane, D.C. Horwood and others were members of the original Post Office technicians team who were transferred from various parts of the Post Office Engineering Department to Dollis Hill to assist in the testing, commissioning, and subse-quently maintenance of the Colossus series of machines, including Colossus I.

Such was the speed at which Colossus could process information (5,000 characters per second on paper tape) that what used to be done in weeks now took hours. Mark I Colossus was upgraded to a Mark II in June 1944, and was working in time for Eisenhower and Montgomery to be sure that Hitler had swallowed the deception campaigns prior to D-Day on 6 June 1944. There were eventually 10 working Colossus machines at Bletchley Park.

The aforementioned report titled *A Technical Description of Colossus I* is an official document written by D.C. Horwood who explains the origin of the report in the introduction:

> This paper has been prepared to honour an undertaking given by GCHQ to prepare, and place in archives, a description of the early electronic computer-like equipment known as Colossus I. The decision to prepare this record followed representations, made by Professor B. Randall of Newcastle University to the Prime Minister in January 1972, which sought the release of information on the machine so that it might be accorded an appropriate place in the history of the development of the modern electronic computer. After careful consideration it was decided that the information requested by Professor Randall could not be released at that time, but that a description of the machine should be prepared so that it would be available if it were to be decided, at some future date, that the information could be released.[15]

This report now forms part of the National Archives. Page 7 of the report provided an outline of description of Colossus, which essentially falls into six basic functions. They are:

a) **Tape reading** – it 'read' the five bit-streams (and the feed holes) of a loop of five 'unit' telegraph tape, simultaneously, at a speed up to 5,000 characters a second.

b) **Internal bit-stream generation** – it generated internally a number of bit-streams in synchronisation with the bit-streams read from the telegraph tape loop.

c) **Combining of bit-streams** – it combined bit-streams selected from those read or generated as described in a) and b) above.

d) **Counting** – it counted the 'marks' or 'spaces' in the bit-streams resulting from the combining described in c) above.

e) **Read-off and printing** – it read-off and printed reference data and the results of the counting described in d) above.

f) **Cycle repetition and advance** – it repeated the cycle of events described in a) to e) above automatically once every revolution of the tape loop. The internally generated bit-stream was changed automatically for each revolution of the tape.

Figure 6.11 shows this diagrammatically.

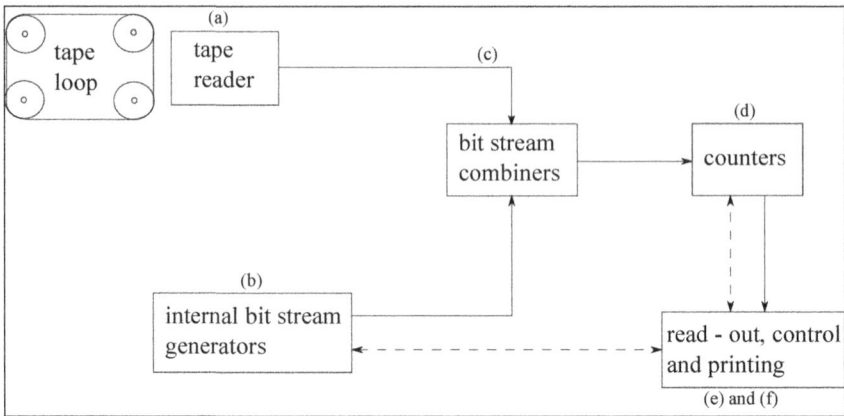

Figure 6.11 Schematic diagram of Colossus. (Author's diagram)

Figure 6.2 detailed some of the Fellows working at Bletchley Park. Of course there were other Fellows working in intelligence departments in other parts of the country. Some of these Fellows were recognised for their war-work. Whilst details are somewhat sketchy, the following is recorded in the Royal Society Biographical Memoirs:

- Claude Elwood Shannon ForMemRS 'was officially responsible for developing fire-control systems for anti-aircraft guns and cryptographic systems. He also developed a mathematical theory of cryptography leading to the system used by Winston Churchill and Franklin D Roosevelt for transoceanic conferences. He met his British counterpart Alan Turing (1943), who was briefly visiting the Bell Laboratories.'[16]
- Dame Miriam Louisa Rothschild FRS 'spent time at the code-breaking centre at Bletchley Park, where she worked on the German 'Enigma' code. She received a government Defence Medal for this work.'[17]
- Dorothy Hill FRS enlisted in the Women's Royal Australian Naval Service as a Third Officer and undertook training in Sydney and Adelaide. Her work included cipher and coding, accepting responsibility for the safety of shipping and communicating with service personnel.'[18]
- Gavin Rylands de Beer FRS 'served at first on the General Staff of the War Office, dealing with Military Intelligence and propaganda. It was during this period that he was elected to Fellowship of the Royal Society – he was believed to the only serving FRS then in the Army. In consultation with the

President of the Royal Society he initiated the Nature Observation (Army) Committee, a joint Committee of the Royal Society and the military arm. The aim was to promote morale by drawing the attention of troops to the phenomena which could be observed and recorded. The habits of wood-pigeon and the fulmar petrel and the migration of birds were among the examples suggested.'[19]

- Hugh Hamshaw Thomas FRS 'joined the R.A.F.V.R. serving first as Station Intelligence Officer at Bomber Command (1939), later posted to the Central Interpretation Unit. When the C.I.U. moved from Wembley to Medmenham, Hamshaw Thomas was put in charge of all R.A.F. specialists in photographic interpretation. In 1943 this included a very detailed analysis of Peenemunde. Retired for health reasons with the rank of Wing Commander (end of 1943).'[20]

- Sir Derek Harold Richard Barton FRS 'worked for the government at a military intelligence unit at Baker Street in London (between 1942 and 1944). Sources close to him understand he was developing invisible inks that could be used on human skin. Sir Derek was awarded the Chemistry Nobel Prize in 1969.'[21]

- Sir Frederick Charles Frank FRS 'was involved in intelligence work, however the references are somewhat obscure.'[22]

Nathaniel Mayer Victor Rothschild FRS (3rd Baron Rothschild), (see figure 6.12), worked in intelligence during WWII. He knew Burgess, Blunt and others who used his London flat as a meeting place. A comment in his Royal Society Biographical Memoir is interesting:

> Early on (in the War), he secured a posting to Intelligence and became Scientific Advisor to MI5 … throughout this time Victor was engaged in counter-sabotage work often of a dangerous kind … His wartime duties involved protecting Sir Winston Churchill from poisoned or exploding substances. He records how Churchill was amused by the experiments conducted to test his cigars for safety, but impatient of any delay, which would give rise to 'irate messages'.[23]

Reginald Victor Jones CH FRS was a little known scientist during the War whose exploits became known over time, especially so following the publication of his memoirs. Jones was made a Scientific Officer at the Air Ministry in 1936, and was seconded for a year to the Admiralty in 1938–39. Figure 6.2 details the work Jones was involved with in Bletchley Park, however, his remit was much more than any involvement in decoding, or the intelligence derived from it. Indeed on 11 June 1940, Jones read a decoded Enigma message which convinced him that the Germans had an intersecting radio beam system for

Figure 6.12 N.M.V. Rothschild. (© Godfrey Argent Studio. Reproduced with permission)

bombing England. Not only did he bring it to the attention of his senior managers in the intelligence community, he attended a special meeting at 10 Downing Street which was called for 21 June when he was just 28.

He advocated the use of echoes from aluminium strips, subsequently code-named 'Window' in 1942, (see figure 6.13), thought to have saved 70 to 80 aircraft in the attack on Hamburg in July 1943 alone. Later he provided evidence to Churchill of V1 and V2 rockets about to be launched on London.

Jones and his group provided vital information concerning the German radar locations on D-Day-1. Air attacks destroyed all but six of the 47 sites. Allied aircraft then scattered 'Window' fooling the Germans into thinking a fleet of destroyers were in the area, thereby masking the approach of the invasion force. The official Despatch concluded: 'These attacks saved the lives of countless soldiers, sailors and airmen on D-Day'. Jones went on to become Director of Intelligence (1946–52), Professor of Natural Philosophy, Aberdeen University (1946–81), Director of Scientific Intelligence, Ministry of Defence (1952–53) and Chairman, Air Defence Working Party (1963–64).[24]

The Royal Society Biographical Memoir for Rendel Sebastian Pease FRS makes the following comment:

> In 1944 he and Thewlis were invited, under conditions of great secrecy, to join a team to implement a countermeasure project, which came to be

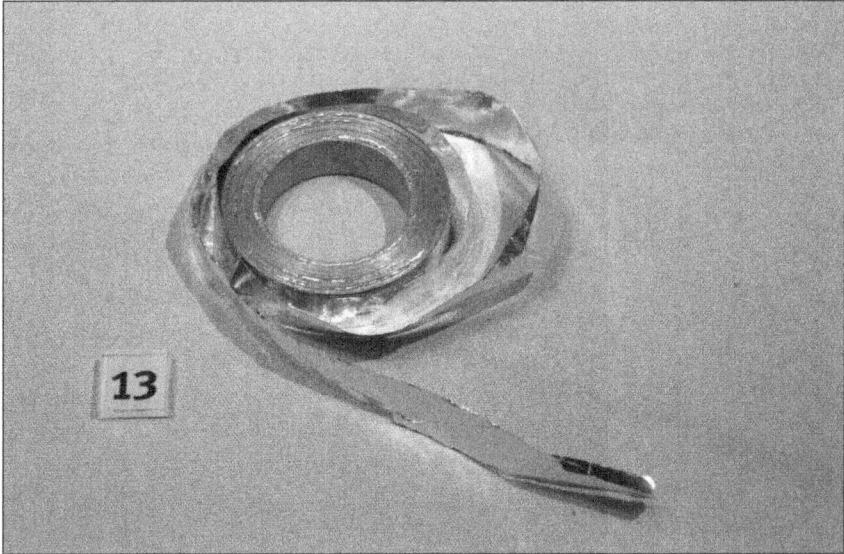

Figure 6.13 Window. (Author's photograph)

known as Operation Glimmer (Hesketh 1999; Holt 2004). The idea had been proposed by Robert Cockburn of the Telecommunications Research Establishment at Malvern to mislead the German radar defences along the French coast in advance of the Normandy landings by simulating a landing at the mouth of the Pas de Calais by means of a curtain of radar-reflecting strips of aluminium foil moving at a suitable 'nautical' speed of 8 knots, laid by a squadron of suitably trained aircraft from 218 Squadron.[25]

Radar

When taught physics at school, one usually is presented with a very simple diagram such as figure 6.14 if there are any discussions concerning radar.

Simplistically, a sender wave is transmitted and bounced back by an object. Measuring the time it takes for the wave to get back to the sender/receiver, and using speed = distance/time, another old classic equation from school, one can find the distance from the sender/receiver to the target. Sending a wave from a different location to the same target will get another distance. Triangulating the two distances from the sending devices (the distance between the two sending signals being known) one can work out the direction as well as the distance.

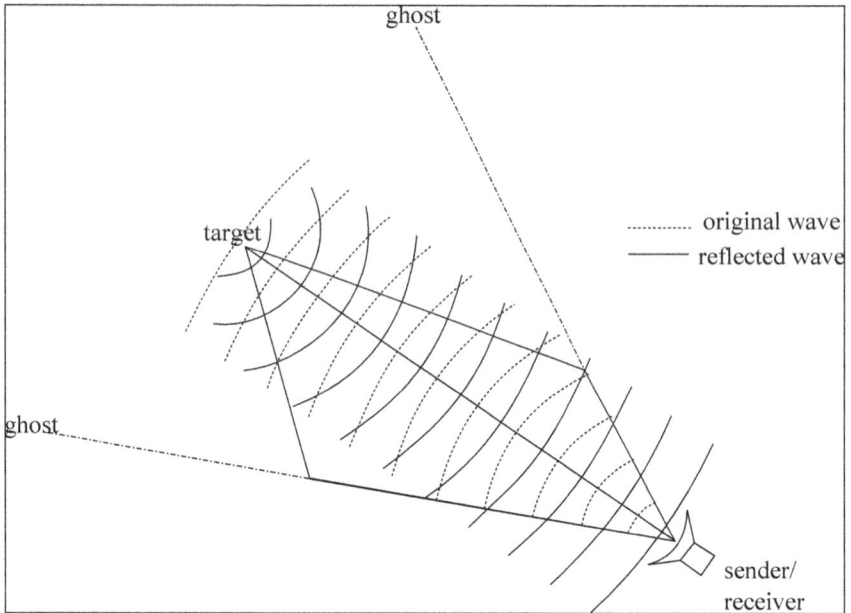

Figure 6.14 Basic principles of radar. (Author's sketch)

This simple model is great for understanding the basics of the technology. Sadly, life is rarely that simple.

What if the target is moving? Furthermore what if the target is in the air and therefore needs three numbers to describe height, distance and speed. Even worse is the problem of a moving sender/receiver, such as a ship or an aircraft. Add to this a distinct lack of computing power, and one begins to appreciate the complexities of radar during the War.

Figure 6.15 is of a memorial to the birth of radar. It details two scientists, namely Sir Robert Alexander Watson-Watt CB FRS and Arnold Wilkins OBE. Wilkins worked with Watson-Watt at the Radio Research Station. The technology was rapidly developed and by the end of the year the range had been extended to 100km. Sir Henry Tizard FRS, then Chairman of the Aeronautical Research Committee, agreed to create the memorial plaque to radar invention radar stations covering the approaches to London, see figure 6.15.

Each Service in the armed forces had its own unique problems to solve. It is easy to understand, therefore, that at the start of the War there was not one central radar development organisation, but several. As time went by there was a consolidation of effort with radar chiefly being developed in Malvern at the Telecommunications Research Establishment (TRE).

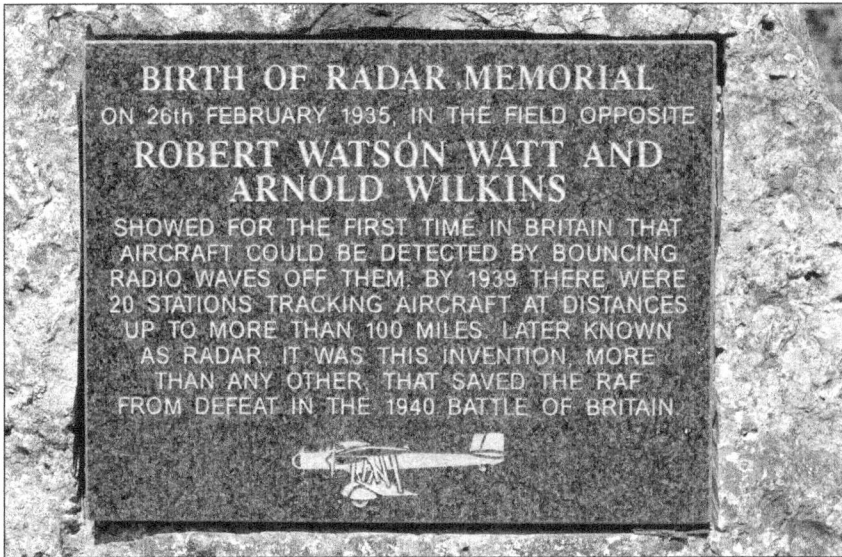

Figure 6.15 The birth of radar. (Author's photograph)

Radar Research Station at Malvern – home of the TRE

The TRE was 'home' to many inventive scientists and engineers whose remit was wider than simply the development of radar. Of course they issued reports such as *RDF. Operations Instructions for working with Jamming* (at that time RDF was the name for radar). This small pamphlet is typical of the short reports issued on a range of subjects. It contains diagrams of the effects of jamming on the radar screen by way of examples, with the following summary:

1) Resolve to keep on working as long as any echo image can be seen.
2) Try the A/CW filter.
3) If (2) is ineffective, switch out the filter and switch on the back-bias.
4) It is possible that the final best results may be obtained by leaving the A/CW filter in operation and turning off the back-biasing, but no hard and fast rule can be laid down about this.
5) Inform G.O.R. as above.
6) If a lull occurs in the main operations make a note (for later reports) of the type of jamming pattern, ie railings, etc. This is only of secondary importance, and the proper use of the GL must not be hindered by worrying about the actual pattern.[26]

The jamming pattern of railings mentioned above was one of six described in the pamphlet which included accompanying diagrams. The descriptors were:

- Inverted echo.
- Tram lines.
- Criss cross.
- Railings.
- Grass.
- Running rabbits.

Early in the War there was a separate organisation called the Air Defence Research and Development Establishment (ADRDE). This organisation and TRE later merged. There were many valid reasons for this merger, including that of avoiding duplication. The following is the start of a report issued whilst ADRDE was a separate organisation:

ADRDE Research Report 236
(Anti-Jamming)
The Effect of Airborne Jamming on SLC Equipment
Authors: J. Briggs, W.S. Elliott and J.T.G. Milne

Summary
Extensive trials are described, of airborne and other jamming, against Mc/s SLC equipments (AA No 2 Mks I–VII), CW, 2 kc/s and noise modulation were used, both with the jammer as the target ('JL Operation') and the target with other than the jammer ('working through').

It was found that a 1 watt jammer at a distance of 5 miles and on a bearing outside the SLC beam, was effective in jamming SLC operation on other targets. The same effect would be produced by a 100 watt jammer at 50 miles distance.

Accuracy of following a jammer with the JL attachment was found to be comparable with normal target following accuracy. Accuracy could, however, be upset by jammer modulation at the SLC aerial switch frequency. The chance of JL following being upset by the presence of more than one jammer is considered small. (Note. This report deals with work done between November 1942 and May 1943. Various interim reports with limited circulation, published at the time of the work, are consolidated).[27]

The following Fellows were at one time or another involved in radar development at either the TRE or ADRDE, see figure 6.16.

Name	Occupation
Andrew (Edward Raymond)	Studied the attenuation of microwave radar signals through gun flashes.[28]
Budden (Kenneth George)	Applications of radar (from 1939).[29]
Callan (Harold Garnet)	Helped to design, install, operate and repair increasingly sophisticated radar detection devices in aeroplanes, landing craft and ground stations.[30]
Cockerell (Sir Christopher Sydney)	Worked on a team building a radar signal detector. The equipment, code-named Bagful, was successfully used to pinpoint German radar stations which were then destroyed before the D-Day landings. It was publicly praised by President Roosevelt as 'one of the highlights of the War'.[31]
Cockroft (Sir John Douglas)	Was onetime Chief Superintendent of the Air Defence Research Development Establishment prior to his involvement in the atomic bomb development.[32]
Curran (Sir Samuel Crowe)	Development of centimetre radar with Sir Alan Hodgkin FRS, Sir Bernard Lovell FRS, Herbert Skinner FRS.[33]
Hey (James Stanley)	Worked under the leader ship of Sir Basil Schonland FRS.[34]
Hill (David Keynes)	His main task was to find how best to make use of the very imprecise data given by the radar sets of that time. Worked under the leader ship of Sir Basil Schonland FRS.[35]
Huxley (Sir Andrew Fielding)	Worked under the leader ship of Sir Basil Schonland FRS, President of the Royal Society (1980–1985).[36]
Jackson (Willis, Baron Jackson of Burnley)	Organised the work of the Signals Research and Development Establishment concerned with the propagation in waveguides and the behaviour of dielectrics at microwave frequencies and detailed investigations of the behaviour of polyethylene, polystyrene and later polytetrafluorethylene as cable insulators.[37]
Kendrew (Sir John Cowdery)	Worked on oscillators and aerials for early radar applications.[38]
Melville (Sir Harry Work)	Scientific Adviser to the Chief Superintendent of Chemical Defence, Ministry of Supply (1940–1943), Superintendent of the Radar Research Station at Malvern (1943–1945).[39]
Mott (Sir Nevill)	Worked under the leader ship of Sir Basil Schonland FRS.[40]
Pippard (Sir Alfred Brian)	Worked on aerials and transmission lines.[41]

Name	Occupation
Schonland (Sir Basil Ferdinand Jamieson)	Deputy to Cockroft (1940–1942) who was Chief Superintendent of the Air Defence Research Development Establishment at Christchurch.[42]
Hodgkin (Sir Alan)	Development of centimetre radar with Sir Samuel Crowe Curran, Sir Bernard Lovell FRS and Herbert Skinner FRS.[43]
Lovell (Sir Bernard)	Development of centimetre radar with Sir Samuel Crowe Curran, Sir Alan Hodgkin FRS and Herbert Skinner FRS.[44]
Skinner (Herbert Wakefield Banks)	Largely responsible for the development of a device which made it possible to send and receive signals on a single aerial. He was also one of the first to work on the detection of submarines by centimetre radar Sir Samuel Crowe Curran FRS, Sir Alan Hodgkin FRS and Sir Bernard Lovell FRS.[45]
Swann (Michael Meredith, Baron Swann of Coln St Denys)	Worked under the leader ship of Sir Basil Schonland FRS.[46]
Wilkes (Sir Maurice Vincent)	Worked under the leadership of Sir Basil Schonland.[47]

Figure 6.16 Fellows involved in radar development.

Admiralty Radar/ASDIC

The following individuals were known to have worked on Admiralty radar or ASDIC:

- Brebis Bleaney FRS 'and the other British born members of the Clarendon Laboratory were employed by the Board of Admiralty as a team working on centimetre-wave devices (from 1st January 1940). Later, Beaney visited the USA, where he was able to interest Raytheon of Waltham, MA, to manufacture their device on a large scale. It has been said that the package he carried was one of the most important to cross the Atlantic in that direction throughout the War.'[48]
- George Porter KT OM FRS, Lord Porter of Luddenham 'served in both the Western Approaches and Mediterranean theatres as a Radar Group Officer (Jan 1942 – Sept 1943) which involved keeping the radar equipment operational on board destroyers therefore called for great improvisational skills; for example, George discovered that the large yellow and black resistors known as 'tigers' could be repaired with boot polish (as reported by Brian Thrush FRS, Cambridge.'[49]

- John Flavell Coales FRS 'was put in charge of all those developments involving centimetric wavelengths (that is, less than 100 cm). He formed a group of about 12 people, and they soon built up a very close collaborative relationship with E.C.S. Megaw and his group at GEC Research Laboratories. To develop an operational radar. A wavelength of 50 cm (600 MHz) was chosen. Every component part had to be worked up from scratch, including modulators, transmitters, receivers, displays, aerials, and even new forms of power unit. As operational requirements were clarified, and specialised effort was organised, the group concentrated on centimetric-wavelength radar for the control of naval guns in range, azimuth and elevation.'[50]

- Maurice Henry Lecorney Pryce FRS 'was directed to work at the Admiralty Signals Establishment (ASE) at Nutbourne in Sussex, one of the wartime centres for the development of radar (in 1941).'[51]

- Sir Alan Hugh Cook FRS 'was sent as a temporary Experimental Officer to the Admiralty Signals Establishment (ASE) at Witley, near Godalming in Surrey, where he was involved in radar research in the Test and Measurement Section.'[52]

- Sir Fred Hoyle FRS 'devised a method of determining the height of enemy aircraft using the radar of major ships of the Royal Navy. They used wavelengths of several metres, which was fine for measuring the bearing of an aircraft, less so for determining its height. Sir Fred realised that the interference lobes of the ship's radars due to sea reflection meant that the ranges at which the aircraft echo faded gave away its height. He produced graphs that were rapidly distributed throughout the Royal Navy.'[53]

- Sir James Woodham Menter FRS 'recruited to work as an Experimental Officer at HM Anti Submarine Experimental Establishment at Fairlie in Ayrshire on the development of various ASDIC devices (1942). He assisted in sea trials of experimental equipment on destroyers on voyages to the Atlantic and Mediterranean.'[54]

- Stanley Keith Runcorn FRS 'started research at the Royal Radar Establishment in 1943, remaining there until the end of the War.'[55]

- Stanley Peat FRS. 'His research group also undertook work for the Admiralty on ASDIC recordings.'[56]

- Thomas Gold FRS. 'One major task was to get at least a rough idea of how the sea state affected radar visibility of conning towers, small ships, and so on, for aircraft mounted radar. His systematic efforts substantially improved our understanding. He next looked at a critical problem for the forthcoming invasion of Nazi-held Europe: how could the large number of landing craft each navigate to its meticulously planned landing spot? It soon became clear that the vessel's radar was the only hope.'[57]

Whilst there was consolidation during the War of radar research and development, there were organisations in other Allied countries staffed by well-known scientists, including:

Australian radar research

The Royal Society Biographical Memoirs records the following for Joseph Lade Pawsey FRS and Sir Bernard Katz FRS respectively.

> When the Radiophysics Laboratory was established in Sydney by C.S.I.R.O. towards the end of 1939 to carry out radar research for the Australian Armed Services, J.L. Pawsley was one of the first staff members to be recruited. He was appointed in London in October 1939 and commenced duty in Sydney on 2nd February 1940, bringing with him a quantity of radio components and test gear for the new laboratory. The first radar equipment to be developed in Australia was a 1.5 metre shore defence and gun laying system (the Sh.D.). In 1943 another group led by Pawsey was formed (together with a mathematical group under Dr John Conrad Jaeger FRS) to study super-refraction phenomena which occurred frequently on the North-West Australian coast.[58]

> Sir Bernard's first military posting (Oct 1942 – Mar 1943) was with a radar unit on Goodenough Island, situated off the north-eastern coast of New Guinea. As a radar Pilot Officer, and later as a Flight Lieutenant, he was in charge of about 20 men running a movable radar unit. The unit surveyed the Japanese planes that were attacking Port Moresby and Darwin, Australia, from their base at Rabaul on the island of New Britain. His second posting (mid-1943 to autumn 1945) was at the Radio-Physics Laboratory at Sydney University, where he was involved in the development of the radar transponder.[59]

There are also entries in the Royal Society Archives for American scientists.

US radar research

- Hans Albrecht Bethe ForMemRS. 'After becoming an American citizen in February 1941, Bethe finally received his clearance to work on classified military projects in December (of the same year). The first such project was linked to the radar that was being developed at the Radiation Laboratory at MIT; Bethe invented the so-called Bethe coupler, a device used to measure the propagation of electromagnetic waves in waveguides. In 1942, while working on the radar project, Bethe participated in a summer study group at Berkeley, organised by Robert Oppenheimer ForMemRS, which led to the creation of the Los Alamos Laboratory in 1943. Bethe played a vital role in the Manhattan Project.'[60]

- John Stuart Foster FRS. 'After the outbreak of World War Two, several physicists at McGill started to work on radar development. Some of this work, particularly that on slotted waveguide antennas by W.H. Watson and his students, was to find its way into battle before the end of the War. The McGill effort, known as the Hush-Hush Lab, flourished until 1944. Foster helped to establish it, but apparently felt, at least in part wrongly, that its efforts might never have a chance to contribute. In any event, in 1941 he went to the newly established Radiation Laboratory of MIT. The Radiation Lab's name was deliberately vague, to cover its activities as the heart of the US effort in radar development. Well-known physicists and engineers, and other scientists from all over the USA were gathered there, some were later to transfer to the nuclear weapons project. Foster's official position was that of scientific liaison officer for the National Research Council of Canada ... he always had a rather informal attitude towards the US-Canadian border and was not upset that what he was smuggling was also officially secret. He remained at MIT until late in 1944 designing microwave antennas.'[61]

- Lyman Spitzer Jr ForMemRS. 'With the outbreak of World War Two, Spitzer took leave from Yale to conduct scientific work in support of the war effort, initially as a member of the Special Studies Group at Columbia, then as director of the Sonar Analysis Group (at the age of 30 years). Radar was the major British technical contribution to the Allied war effort. Although Spitzer was always modest about the development, sonar along with the much more recognised A-bomb effort was one of the decisive technical contributions to the US War machine.'[62]

UK radar Manufacture

Thomas Edward Allibone FRS worked in the Research Department of the Central Electricity Board. He was heavily involved in the manufacture of the following radar components:

i)   A three-electrode valve, the 'trigatron'.
ii)   Direct-current (d.c.) supplies.
iii) Magnetrons (operating at 70 kV for the Admiralty and 8 kV for airborne use).
iv)   Triggering mechanisms needed to produce the square-wave pulses to the emitters.[63]

Finally, there are some technologies defying categorisation. The following are two examples of active and fertile minds solving War related problems.

Irvin Langmuir ForMemRS studied methods of:

- Producing 'white' smoke screens.
- The capacity of gas masks to filter out particles of a given size.
- The production of rain by a chain reaction in cumulus clouds at temperatures above freezing.
- The production of snow from supercooled clouds.
- The control of precipitation from cumulus clouds by various seeding techniques including:
  - i)   Silver iodide.
  - ii)   Sodium chloride.
  - iii) Solid carbon dioxide.[64]

John Walter Ryde FRS was attracted to the problem of fog dispersal on airfields by the liberation of large quantities of heated air – a project known as FIDO. It used a vertical air-curtain which entrained air from either side and could be made to act as a sucker to draw warm air across the runway so as to cope with most conditions of cross-wind. This scheme was in the end not used because the fuel position became so much easier.[65]

Interestingly there was an informal meeting on Non-Chemical Smokes, the second meeting of which was held on 8 April 1943. William Alexander Waters FRS was amongst the attendees. Item 7 from the minutes reads:

> An advance copy of the minute of the Smoke Co-ordinating meeting held on the 31st March does not affect the terms of reference of this Committee, except in referring to the development and improvement of Daylight Haslars. This is on a lower priority than the problem listed in the terms of reference and does not therefore affect this Committee immediately.[66]

There was interest in smoke from a number of interested parties.

7 Armaments

Most of the armaments used in the War might be classed as conventional. That is to say that some sort of casing enclosed an explosive, set off with a charge either in the barrel of a weapon or by contact, as in a bomb dropped from an aircraft. Conventional armaments have often evolved from much earlier weapons, the charge, method of propulsion/detonation and the means of deployment changing with every generation of technology. However, this War more than any other, saw the deployment of weapons not seen before.

The use of atomic fission in warfare was revolutionary. One could argue that the 'bouncing bomb' and the raid on the Ruhr dams was novel. Actually, Lord Nelson used the technique of bouncing cannon balls on the sea many years before the bouncing bomb was built. Nevertheless, for the purposes of these discussions, these two types of armament will be treated separately, compared with conventional weapons. Additionally the development of gas weapons was not novel, one only has to read accounts of World War One, nevertheless this type of armament will also be treated separately.

As the War moved into the 1940s, the need for more complex instruments of War and the munitions to fire from them must have strained an already overburdened system. Accordingly H.J. Gough FRS, the Director of Scientific Development for the War Department and Ministry of Supply called a meeting, leading to the Guy Committee on Armaments Development. Gough wrote this letter:

> You will recall the meeting with the Minister and Dr Guy on 14.5.42 at which both of us were present.
>
> The Minister decided that the Scientific Assessors could now be selected in consultation with Dr Guy and I was instructed to take the necessary action. Yesterday I held an informal meeting at which the following were present:

Sir Henry Guy FRS	Chairman of Committee
Sir Edward Appleton FRS	Secretary DSIR and Member of Committee
Sir David Pye FRS	Director of Scientific Research, MAP
Mr C.S. Wright	Director of Scientific Research, Admiralty was also invited to attend, but was unable to do so

We made a very careful review of a very large number of names of scientists, including all the senior scientists connected with the Ministry of Supply Research and Development, on the Scientific Advisory Council and its Committees and consultants to the Ordnance Board, etc.; we also considered a number of names of scientists eminent in industry for organisation and conduct of research and development work.

As a result it was unanimously agreed to nominate the following names:

- Professor Sir Robert Robinson FRS (Waynflete Professor of Chemistry, University of Oxford) 117, Banbury Road, Oxford.
- Professor Sir John Lennard-Jones FRS (Plummer Professor of Theoretical Chemistry, University of Cambridge) Middlefield, Huntingdon Road, Cambridge.
- Sir Clifford Paterson FRS (Director of Research Laboratories, General Electric Company, Wembley) Waldringfield, Oxley, Herts.
- W.T. Griffiths (Manager, Research and Development, Mond Nickel Co. Ltd.) The Mond Nickel Co. Ltd., Grosvenor House, Park Lane, W1.

I should be grateful if you would formally approve these nominations so that invitations to serve can be at once sent out. I may mention that the first meeting of the Committee will be held on Thursday of this week, and we are anxious to assemble the assessors as soon as possible.

If you approve the nominations, please return this paper to Mr Ivimy so that he may prepare the draft letters of invitation.

H.J. Gough FRS
19/5/42[1]

Born on 15 June 1887 Guy joined the Technical Staff of British Westinghouse in 1910 and from 1919 to 1942 was Chief Engineer in the Mechanical Department of Metropolitan Vickers Company. Guy was a:

- Member of the Advisory Council for Research and Development in the Ministry of Supply (1939–43 and 1943 onwards).
- Chairman of the Gun Design Committee.
- Chairman of the Committee on Armaments Development (1942).

- Chairman of the Static Detonation Committee (1943).
- Chairman of the Aircraft Armament Development, MAP (1945).
- Chairman of the Committee on Technical Organisation of the Army (1944).
- Member of the Advisory Council to the Committee of the Privy Council for Scientific and Industrial Research (1944 onwards).
- Member of the Mechanical Engineering Advisory Committee for the Ministry of Labour (1942 onwards).
- Chairman of the Armaments Development Board, Ministry of Supply and of Aircraft Production (1945–47).
- Chairman of the Committee on Essential Requirement of Mechanical Engineering Research DSIR (1945–46).[2]

Sir Henry was elected a Fellow of the Royal Society in 1936.

The Guy Committee's actions were both far ranging and comprehensive, the full Guy report appears as Appendix VIII.

The following was an initial reaction to the report:

Figure 7.1 Sir Henry Lewis Guy. (© Godfrey Argent Studio. Reproduced with permission)

Secretary

The Guy Committee report had been circulated to the Chairman of the Supply Council, CGMP, CGRD, SSO, DGM, DGW and DGGA for comments. Copies have also been sent with a covering letter to the Admiralty, War Office, Air Ministry, Ministry of Aircraft Production inviting comments and telling them that a conference will shortly be called to discuss the line of action. Copies have also been sent for information to the Minister of Production, DSIR and the Treasury.

The Minister approved the suggestion that the conference to discuss the line of action on the report should be convened by Sir Harold Brown. Departments have been informed accordingly. It is important that before the conference takes place he should also see the departmental remarks from his Ministry.

We are due to comment on the report from the point of view of the Secretariat. Mr Dobbie-Bateman's remarks are attached. I have sent a copy of the report separately to D.of E. and asked him to let you have direct any comments from the Establishment point of view.

Generally, I think the proposals relating to the Research Department and Armaments Design Department are satisfactory and should, subject to details, be accepted. I agree with Mr Dobbie-Bateman that the Committee's recommendation about the strengthening of the Design Department are particularly welcome. It is quite clear that we cannot afford to leave this important department staffed on its present inadequate basis.

As regards the Ordnance Board, the proposals of the Guy Committee are quite obviously a series of compromises which will leave the Board substantially in its present position. I do not think that this is satisfactory. In my view the Ordnance Board requires to be radically reformed. I believe there is considerable opinion in favour of abolishing it altogether. I would not go so far as this – if only because a body of this kind does not exist as long as the Ordnance Board has existed without having a reason for existence; but I am quite sure its wings should be clipped, its size reduced, and its membership built up on an entirely different basis. There should be no room in wartime, in the scheme of the munitions design development and production, for a body which gives a home to a number of officers who have been regarded as finished for purposes of other technical deployments. As pointed out by Mr Dobbie-Bateman's note, the size of the Board alone is fantastic under present circumstances.

At meeting after meeting during the period of the Rearmament Program and ever since the War started we have heard that matters of importance from the point of view of getting on with production are held up pending consideration by the Ordnance Board. Not only has this been the case, but the Board has been a well known centre of reaction and only recently considerable embarrassment has been caused by its criticism of decisions which have been properly taken in the interests of facilitating production under wartime condition. In this connection it is important to note that there is a view held in the War Office that the Board provides a safeguard for Service interests against the Ministry of Supply. Also to recall that the Board is Ministry of Supply responsibility.

Radical reform of the Board may be a difficult problem and I do not suggest that it could be undertaken immediately. On the other hand I do not think we ought to let it be thought that we accepted the milk and water recommendations of the Guy Committee as in any way providing a real solution. I am quite sure that until the problem of the Ordnance Board is tackled, all other alterations will prove to be nothing more than adjustments. If it is to be tackled we need a very small authoritative Committee under someone like Lord Weir.

If the Ordnance Board is maintained, I cannot see what advantage is to be obtained from setting up another Inter-Service Committee. I think there should be a much better justification for a new Inter-Services Committee recommended in paragraph 15 of the report before it is accepted. As I see the proposal at present it will only be another obstacle in the way of getting things done quickly.

When Sir Harold Brown calls the interdepartmental conference, the practical line of progress will probably be to reach agreement on nominations for the Civilian Heads of the Design Department and Research Department. I understand that opinion has crystallised in favour of the following appointments:

To be Chief Engineer and Superintendent of Armament Design	Dr Guy
To be senior Principal Design Engineer	Mr Hillier
To be Chief Superintendent of Armament Research	Professor Leonard Jones

It is perhaps a little unfortunate that these three nominations are the Chairman and two of the Assessors of this Committee, but if they are the right appointments that need not worry us. I believe that Professor Leonard Jones is well suited for the appointment. Mr Hillier has already done a great deal of work in connection with the design of armaments and it is quite clear that we ought to get him fully into the work; he is not regarded as a good selection for the top job, but as an excellent number two.

As regards Dr Guy's appointment, I am bound to say that I am very doubtful whether it is likely to achieve the desired result. His standing is a very good; no one can deny his qualifications, but on the Guy Committee he has proved himself adept at dealing with the complicated interplay of Service and Scientific opinion which centres around 'Woolwich Organisations'; and I very much fear that in him we shall get perhaps a better controversialist than head of a Design Department. The Service Departments may very well fasten on this aspect and fail to reach agreement on this appointment. On the other hand, if they do agree, there is nothing to be done for the very simple reason that there is no other satisfactory candidate in view.

At the back of this discussion there is always the Service Departments very reluctant acceptance of the present position whereby the Technical Establishments are administered by the Ministry of Supply. If things do not go right or if there is any undue appearance of neglect of the Services point of view, this issue is bound to be reopened. It is one which would cause grave difficulty. For this reason alone, it is most important that both changes of organisation and of staffing should be on sound lines which the Service Departments cannot contest.

It was worth reviewing this letter in its entirety as it sets the tone and allows the inter-departmental politics of the day to help shape our understanding of the effect of the Guy Report.

The following are a few paragraphs from the first minutes from a Committee appointed to consider the report which are stored in the same folder in the National Archives.

Notes of the Interdepartmental Meeting held in Room 241 Shell Mex House on Monday 7th September 1942, to consider action to be taken on the recommendations of the Guy Committee's Report on Armaments Development

Present:

Engineer Vice-Admiral Sir Harold Brown	(in the Chair)
Vice-Admiral W.F. Wake-Walker	Controller of the Navy
Rcal-Admiral O. Bevir	Director of Naval Ordnance
Lt-General R.M. Weeks	D.C.I.G.S., War Office
Air Marshal F.J. Linnell	C.R.D., Ministry of Aircraft Production
Air-Commodore G.A.H. Pidcock Aircraft Production	D. Arm. D., Ministry of
Major-General E.M.C. Clarke	D. of A., Ministry of Supply
Mr Graham Cunningham	C.G.M.P., Ministry of Supply

1. The Chairman said he had been directed by the Minister of Supply to call the meeting with a view to considering what action should be taken on the recommendations made in the Report of the Committee on Armaments Development presided over by Sir Henry Guy FRS. He therefore asked the meeting, firstly, whether the recommendations of the Committee were agreed to and, secondly, what action considered should be taken.

    It seemed to him that the most urgent question was the appointment of Heads of the Design Department and the Research Department (C.E.A.D. and C.S.A.R.). He thought that the staffing of these two Departments, in so far as it needed to increase, must be dealt with as recommended in the Report, after the above two officers are appointed.

    The Controller said he agreed with that view. The Chairman asked first of all if it was generally agreed that the recommendations of the Report, so far as the Design Department and the Research Department were concerned, should be adopted.

2. The Controller said there was one point on which he was not quite clear (para 18 (ix)) – the co-ordination by A.R.D. of requests from the various Services.

    The Chairman thought that in so far as these could not be co-ordinated by C.S.A.R., any questions of priority of work were settled the priority given by the ordering authorities (D.N.O., D. of A. and D. Arm. D.), and were questions of inter-service priority arose, they could be settled by the Inter-Service Committee recommended in paragraph 15, with appeal to higher level as required.

Further meetings were held, the next on 2 September 1942. A lot of effort took place to fully understand the implications of the report and which parts to implement in the first instance.

These meetings triggered a small flurry of activity with several Committees formed to assess the future of small armaments research and development. For example:

Ministry of Supply
Committee on Armaments Development
Constitution and Membership

Constitution
Chairman   H.L. Guy FRS
Members
Engineer Vice Admiral Sir George Preece (Admiralty)
Captain O. Bevir Director of Naval Ordnance, Admiralty

Air-Commodore S.A.H. Pidock Director of Armaments Development, Ministry of Aircraft Production
Air-Commodore P. Ruskinson President of the Air Armaments Board Ministry of Supply
H.J. Gough FRS Deputy Comptroller-General of Research and Development, Ministry of Supply
A.F. Dobbie-Bateman Ministry of Supply
Major General L.D. Hickes Director of Staff Duties War Office
A representative of the Scientific Advisory Committee
Secretary   J.W.L. Ivimy Ministry of Supply

Terms of reference

To review the machinery for the conduct of research, design and experimental work in connection with the development of guns, small arms, ammunition and cognate stores; to consider, in particular, the functions and organisation of the Ordnance Board, the Armaments Design Department and the Research Department; and to make recommendations.[3]

Figure 7.2 is an organisation showing their recommendations.

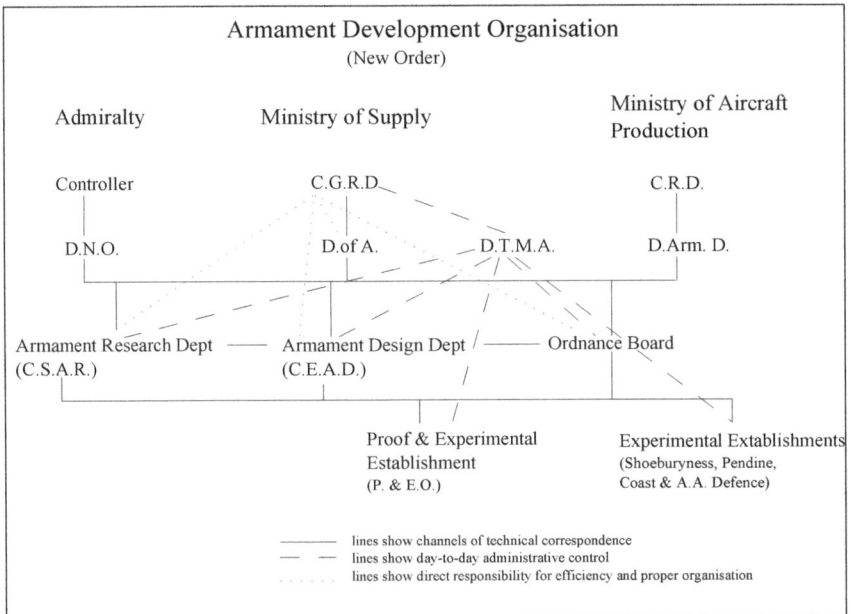

Figure 7.2 The New Order for the Armaments Development Organisation.
(Author's diagram)

Of course producing the armaments on time was only part of the story. Fuses were needed in equal amounts. The first few paragraphs of a meeting held on the 27 April 1943 shows the vast quantities needed.

Present:

Mr Duncan Sandys MP in the Chair

Rear Admiral Bevir RN	DNO	Dr Gough FRS	DGSRD
Captain Woolerton RN	DNO	Mr McLay	DGAP
Commander Madden RN	DGD	Dr Paris	CPRSD
Mr Huxley	DGD	Mr Thorpe	DAI
Professor Cockroft FRS	ADRDE	Dr Vick	SR1
Lt.Colonel Post	M.A./Parliamentary Secretary		

1. M32 (USA) fuse.
    a) D.N.O. said that the Admiralty were hoping to obtain 15,000 of these fuses for 5.25 guns and 25,000 for 4.5 guns from the United States. The Admiralty would, he stated, obtain a firm offer and dates of delivery from America as a matter of urgency. The employment of the M32 would demand a new design for both types of guns.
    b) D.N.O. stated that the designs of the new shell required for use with the M32 fuse would be clear by the time an answer could be received from America, and that they would in any case start production of 5.25in shell as soon as design was clear.
    c) D.N.O. stressed the difficulties arising from the employment of the M32 under service conditions. It would, he said, be of great help if batteries for replacement could be supplied from production in this country as well as from America.[4]

The logistics of armament supplies was further complicated as 24 fuse suppliers were under contract just for the supply of fuses 712, 715 and 716. The rationale for such a large number of firms was that not all suppliers provided all of the parts for each fuse in order to keep the overall design as secret as possible. Of course any changes to the armaments through a change to the organisation of armament research, such as the speed of introduction of a new design, may have impacted other parts of the supply chain. This issue of fuses is used here by way of an example of the complexity of the overall armaments supply chain.

The problems relating to the Ordnance Board continued after the publication of the New Order organisation chart. The functions of the Board were earlier documented as being:

i.   To undertake such investigations as were required of it by the Admiralty, War Office or Air Ministry in regard to the following matters.

   a)   Questions relating to improvements in or suitability of steel or other materials used in the construction of ordnance, small arms or ammunition.

   b)   New designs, or important modifications to existing designs of guns, breach mechanisms, small arms and ammunition including anti-tank mines, grenades and bombs.

   c)   Questions relating to propellants and explosives.

   d)   Questions relating to armour.

ii. Within their discretion to initiate any question involving research, experiment and investigation with the construction and efficiency of guns, small arms, ammunition, bombs and explosives for the three Services. To originate the consideration of questions affecting progress in artillery science and improvements in design and material and to bring to the notice of the Directors concerned the directions in which it was considered that investigations were desirable.

iii. To investigate all questions in regard to internal and external ballistics.

iv. To investigate accidents to or failures with guns, small arms, ammunition, explosives and bombs.

v.   To undertake investigations for private firms referred to them by the Service Ministries.[5]

This level of responsibility for such a significant part of the armaments supply chain would need careful planning before changes were implemented and outside of the scope of this discussion.[6] Of course there were Fellows of the Royal Society involved in armaments research. For example the following reports were written by A.R. Ubbelohde FRS of the Armament Research Department.[7]

1)   Mechanism of damage in steel targets – the critical change in C.T. damage to armour plate with change in area of attack with T. Nash.

2)   The Structure of Munroe Jets. Part 1. Effect of stand-off on depth of penetration with G.E. Roberts.

3)   Note on the protective power of INDASCO plastic against Munroe Jets with G.E. Roberts.

4) Target damage by Munroe Jets. Effect of target strength on crater volumes with G.E. Roberts.

5) Target damage by Munroe Jets. Comparative damage in face hardened armour plates, home-armour plate and mild steel with G.E. Roberts.

6) Note on the Bakelite charge cases for hallow charges.

7) Target damage by Munroe Jets. Protection of steel targets by non-metallic substances, with special reference to oxidising agents. Interim report with G.E. Roberts.

8) Terminology of Munroe Jets.

There are many other reports written by A.R. Ubbelohde FRS. Born in Antwerp in 1907, Ubbelohde wrote over 400 research papers during his career. He wrote on various topics including chemical thermodynamics, combustion, explosions and detonations, ionic melts, graphite and intercalation compounds. He was also a pig farmer and an expert on Chinese, ceramics wine etc.[8]

The Royal Society Biographical Memoir for Edward Armand Guggenheim FRS contains the following, demonstrating that even with all of the planning on armaments research and development there is still room for one more department.

> By the middle of 1940 a new group began to grow in the Admiralty. It was responsible for the development of Miscellaneous Weapons and Devices, with its staff known as the 'Wheezers and Dodgers'. It was concerned with a wide range of developments, the principle ones being anti-aircraft, anti-submarine and beach-head assault weapons. Some of the specific parts he played in DMW and D are well documented in a book by Gerald Pawle, The Secret War, 1939–45 (Harrap and Co 1956) … The landing in France took the pressure off DMW and D and the Admiralty began to look ahead.[9]

Other Fellows involved in conventional armaments include:

Name	Activity
Bartlett (Maurice Stevenson)	Worked at the Projectile Development establishment of the Ministry of Supply on the theoretical effectiveness of rocket weapons.[10]
Bates (Sir David Robert)	Sent to the Admiralty Research Laboratory at Teddington as one of a group of young scientists, the others being Francis Crick FRS, Sir John Gunn and Richard Buckingham, to assist Massey in developing protective measures against the enormous threat of enemy magnetic mines to ships.[11]

Name	Activity
Boys (Samuel Francis)	Joined the Ballistics Branch of the Armaments Research Department in the Woolwich Arsenal section where he conducted a small research group. By the end of the War he was a Senior Experimental Officer and head of a section.[12]
Christopherson (Sir Derman Guy)	Appointed Scientific Officer in the Ministry of Home Security in the research and experimental department under Sir Reginald Stradling FRS. Worked on the effects of explosives on buildings, fire-fighting and shelters (until 1946).[13]
Denbigh (Kenneth George)	Joined the laboratories of the Royal Ordnance Factory at Bridgwater in Somerset. The plant was producing a new explosive called RDX.[14]
Hewitt (Eric John)	Appointed to the Royal Ordnance Factory, Pembrey.[15]
Howarth (Leslie)	Recruited to work for the External Ballistics Department of the Ordnance Board (1939–1942), Armament Research Department (1942–1945).[16]
Irwin (George Rankin)	Head of the Ballistics Branch at the NRL worked on improvements in armour and damage resistant structures and later to laminated body armour made from non-metallic woven cloth.[17]
Kendall (David George)	Posted to a group of mathematicians working in the Projectile Development Establishment (PDE) at Aberforth. The group was initially led Sir William Cook FRS and later by Louis Rosenhead FRS.[18]
Milne (Edward Arthur)	Employed by the Ministry of Supply and classed as a Key Scientist (1939–1944). He was an Associate Member of the Ordnance Board in 1939 and again from 1943 onwards and was Resident Adviser Ordnance Board 1943–1944. His work dealt with ballistics, the analysis of armour-piercing experiments, rockets, sound-ranging and the optimum distribution of AA guns.[19]
Mitchell (John Wesley)	Joined the Armament Research Department of the Ministry of Supply at the Woolwich Arsenal as a Scientific Officer (June 1940) working on distortions of brass cartridge cases when fired in M2 Browning guns. He was moved to the Armaments Research Establishment at Fort Halstead where he collaborated in studies of shockwave interactions.[20]
Morrogh (Henton)	Chief Investigations Officer for The British Cast Iron Research Association (from 1942/3).[21]
Sneddon (Ian Naismith)	Joined the Ministry of Supply, as a Junior Scientific Officer (up to the end of hostilities in 1945). Seconded to the Cavendish Laboratory at Cambridge to work with W.R. Dean and J.W. Harding on a phenomenological theory of armour penetration.[22]

Name	Activity
Sugden (Samuel)	Appointed Superintendent of Explosives Research on 12th October 1942. Resigned in June 1943 and soon after became Scientific Advisor to the USA 8th Army Air Force in Great Britain.[23]
Travers (Morris William)	Joined a research group under Professor W. E. Garner and Professor E.L. Hirst at the University of Bristol working for the Armament Research Department of the Ministry of Supply. He also became a consultant to the Explosives Section of the Ministry which he joined in February 1940, a few days before his sixty-eighth birthday.[24]
Tukey (John Wilder)	Worked in Fire Control Research (gun and artillery control, range finding etc.) in New Bedford, Massachusetts (from 1941).[25]

Figure 7.3 Some Fellows who worked in armaments development.

Atomic Bomb

It is perhaps remarkable that the fundamental scientific principle needed to even consider the development of an atomic bomb, that of atomic fission, was only discovered just prior to the start of the War. Within the lifetime of the War not only did scientists understand the underlying principles, they developed the means to turn this basic science into a bomb of previously incalculable ferocity. Listed below is the timeline of British atomic energy development.

History of the British Atomic Energy Project During War of 1939–1945

January 1939	The basic scientific discovery of fission in Germany and Denmark.
April 1940	M.A.U.D. Committee set up under the Air Ministry, with Sir George Thomson FRS as Chairman, to examine the whole problem and to co-ordinate work in progress.
June 1940	Arrival in U.K. of Dr Halban and Dr Kowarski, with the French stock of heavy water. [Reference the Suffolk/Golding Mission mentioned in Chapter 1.]
April 1941	Visit of Professor Bainbridge from U.S.
July 1941	Visit of Professor Lauritsen from U.S.
15 July 1941	M.A.U.D. Committee reports recommending prosecution of the project on a large scale, and this report referred to the Hankey Committee.

September 1941	Sir John Anderson FRS took over personal responsibility for the project, and set up a Consultative Council.
October 1941	Directorate of Tube Alloys set up under Sir Wallace Akers FRS as part of D.S.I.R., the Lord President of the Council at the time being Sir John Anderson FRS, and the Secretary of D.S.I.R. being Sir Edward Appleton FRS.
11 October 1941	Approach by Roosevelt to Churchill for joint work on the atomic bomb.
November 1941	Visit to U.K. of an American Mission (Pegram and Harold Urey ForMemRS) representing the U.S. project.
February 1942	Return visit of British Mission to U.S. (Sir Wallace Akers FRS, Halban, Sir Rudolf Peierls FRS and Sir Franz Simon FRS).
June 1942	M.W. Perrin visited Canada to discuss (a) supply of raw material; (b) plans for Halban to lead a team to work in Canada.
September 1942	Halban visited Canada in connection with the proposed Montreal Laboratory.
October 1942	Report of the U.S. Committee under J.B. Conant ForMemRS handed to Sir Wallace Akers FRS in Washington.
January 1943	The equipment and team working under Dr Halban at Cambridge were transferred to Montreal.
May 1943	Churchill met Roosevelt in Washington.
July 1943	Sir Marcus Oliphant FRS sent to U.S. to lead a British team working on electromagnetic method.
August 1943	Quebec Agreement, setting up Combined Policy Committee, initialled. Sir James Chadwick FRS appointed Scientific Adviser to the British members of C.P.C.
October 1943	Professor Bohr ForMemRS reached U.K.
3 December 1943	British Mission of 16 members, led by Sir Wallace Akers FRS, arrived in U.S. and remained till end of January.
January 1944	Cockcroft took over the Montreal Laboratory.
March 1944	At this time, there were about 30 senior British Scientists in the U.S., working at Los Alamos, Berkeley, New York or Washington on Weapon Development, Electromagnetic Separation, Diffusion Separation of Isotopes or on general problems. Other staff went later.

13 June 1944	Declaration of Trust signed for common holding of uranium ores.
February 1945	About 50 British scientists now working with U.S. project.
16 July 1945	Trinity Test at Alamogrado, New Mexico.
6 August 1945	Bomb dropped on Hiroshima.[26]

For completeness, the History of Anglo-American Relations in Atomic Energy up to the Quebec Agreement appears as Appendix IX.

The above timeline shows the responsibility of the atomic bomb effort for the UK falling to Sir John Anderson FRS. This followed a letter from Lord Hankey FRS to Sir John Anderson FRS.

25th September 1941
My dear Anderson,

I enclose herewith a copy of a report by the secret Defence Services Panel of the Scientific Advisory Committee on the projects denoted by the letters M.A.U.D.

The Committee have made a formal recommendation that research and development relating to the power project should be placed under the Secretary of Scientific and Industrial Research (paragraph 4).

My Committee are of the opinion that research and development on the bomb project, notwithstanding that it deals with a War weapon, should also be entrusted to the Secretary of Scientific and Industrial Research. The only reason why we have not included a formal recommendation to this effect is that Sir Edward Appleton was reluctant to associate himself with a recommendation for transferring, to his own Department, a responsibility that has thitherto been so ably discharged by the Ministry of Aircraft Production. I understand, however, that the Secretary of Scientific and Industrial Research is quite willing to undertake this responsibility if Ministers decide to entrust it to him.

My Committee has recommended that no decision should be taken as to proceeding with the later stages of the bomb project until a pilot plant has been constructed and the various other investigations enumerated in the report have been carried out (paragraph 17).

I should mention also that my Committee attach great importance to proceeding at a very early stage with the investigation of the fusing mechanism (paragraphs 8 and 9). I think I am right in saying that we incline to the view that this matter is liable to involve greater difficulties than have hitherto been expected. The Committee would like to communicate certain technical observations, which are not mentioned in the report, to Dr Fergusson of the Research Department at Woolwich as soon as this work is authorised.

The Committee have devoted a great deal of time to this matter and have held seven meetings, nearly all of them of several hours duration. I feel sure, however, that my colleagues are as anxious as I am to render any possible further assistance in this important matter, either now or at any time, for example, if deemed appropriate, when the time comes to take the decision about the later stages of the bomb project.

In conclusion I should like to mention the invaluable help that has been rendered to the Committee in this matter, as always, by Mr Rickett the Secretary.

Yours sincerely
Hankey[27]

The British atomic development program, known as Tube Alloys, started in late 1941. As one might imagine, some of the best scientists in the country took part in these early discussions, including Sir James Chadwick FRS the 1935 Physics Nobel Laureate. The first meeting of the Tube Alloys Project Technical Committee set the scene:

Tube Alloys Project
Minutes of 1st meeting of Technical Committee
16 Old Queen Street, 6th November 1941

Present: Sir Wallace Akers FRS   Chairman
           Sir James Chadwick FRS
           Sir Franz Simon FRS
           Dr H. Halban
           Sir Rudolf Peierls FRS
           Dr R.E. Slade
           Mr. M.W. Perrin   Secretary
           Mr. J.F. Jackson   Secretary

1.   Constitution of Technical Committee:
     The Chairman explained the new arrangements by which all work on the utilisation of nuclear energy will be done under Government control through the D.S.I.R.
     It was pointed out that the M.A.U.D. Policy and Technical Committees had not been formally advised of the new arrangements and this had led to misunderstandings which should be cleared up if full co-operation was to be obtained.

The Technical Committee recommends that a letter should be sent by the Ministry of Aircraft Production to the members of the M.A.U.D. Committees reporting the Government's decision and informing them that this carries with it the dissolution of the Committees.

It is further recommended that all members of the M.A.U.D. Policy Committee and certain members of the M.A.U.D. Technical Committee, whose names would be put forward by Professor Chadwick, should be advised of the method by which the work will be carried on.

Control of Patents:

Discussions with Professors Pegram and Urey ForMemRS, the representatives of the N.D.R.C. of the United States, have convinced the Committee that the closest co-operation in Research between the U.K. and U.S.A. is essential if the development of nuclear energy in all its aspects is to be carried out as quickly as possible.

The Committee is of the opinion that such co-operation can be achieved only if the Governments come to some arrangement in connection with patents.

This seems necessary, also, to prevent the creation of a hopelessly complicated patent position which will otherwise be inevitable.

Any arrangement will have to take account of the patents already held by Dr Halban and his co-workers, who intend to devote a great part of any royalties earned therefrom to the subsidising of scientific research in France and the U.K.[28]

By the third meeting, the project was well and truly underway, even though it was only a few weeks after the project launch.

Tube Alloys Project
Minutes of 3rd meeting of Technical Committee
16 Old Queen Street, 13th January 1942

Present: Sir Wallace Akers FRS   Chairman
             Sir James Chadwick FRS
             Sir Franz Simon FRS
             Dr H. Halban
             Sir Rudolf Peierls FRS
             Dr R.E. Slade
             Mr. M.W. Perrin   Secretary
             Mr. J.F. Jackson   Secretary

15. Minutes of Second Meeting:

The Minutes of the 2nd Meeting were confirmed.

16. Work at Imperial College:

The Chairman reported that, in accordance with the recommendation of Minute 8, Dr Moon had visited Imperial College and that he agreed with the decision of the Technical Committee that work on the separation machine should be suspended. Official notice to this effect had been given to the College Authorities, who are   prepared to consider specific requests for the services of Dr Stephens or of the mechanics in connection with other problems. The decision had been discussed with Dr Stephens.

17. Control of Patents:

Arising from Minutes 2 and 7(b), the Chairman reported that Mr. Hollins had prepared two draft notes dealing –

(a) With the wartime positions, in which effective control is possible in this country and can probably be arranged in the U.S.A., and

(b) With post-war developments, which can be considered more carefully and probably require legislation.

The draft note on wartime control was discussed by the Committee and the following action, with regard to publication, was recommended. The Ministry of Information should be asked to prohibit articles, in the non-technical press, describing the construction, composition or method of operation of bombs or projectiles. The Editors of the following journals:

Nature

Proceedings of the Royal Society

Philosophical Magazines

Proceedings of the Physical Society

Cambridge Philosophical Society

Transactions of the Faraday Society

should be asked to submit all communications, before publication, to the Hankey Committee or to an individual appointed by Lord Hankey. No papers bearing on the Tube Alloys field should be published without the written consent of D.S.I.R.

18. Minutes of Chemical Panel (1st Meeting)

The Minutes of the 1st meeting were noted by the Committee.

The following action was reported, arising out of these minutes:

(a) Two tons of U_3O_8 (98% pure) had been ordered from Eldorado through the National Research Council of Canada.

(b) Stocks of uranium residues in Portugal, owned by the U.K.

Commercial Corporation, will be shipped to this country.

(c) Arrangements had been made for Dr London to start work at Winnington on the electrolytic method of isotope separation.

19. Stocks of Uranium Oxide:

It was pointed out that stocks of uranium in this country are small and that it would be advisable to buy a further supply from Canada. It was suggested that 25 tons should be ordered but, as the price is 3 dollars per pound, a considered argument must be prepared for submission to the Treasury.

It was agreed that the question of stocks of uranium should be referred to the Chemical Panel.

20. Naturalisation:

The Committee recommended that the cases of certain aliens who are engaged on Tube Alloys work should be specially considered. The following have the necessary qualifications for naturalisation and wish to be naturalised. They are known well to members of the Technical Committee:

Dr K. Fuchs
Dr Heinz London FRS
Dr S. Bauer
Dr A.F. Freundlich
Dr N. Kemmer FRS

The Chairman agreed to raise this question at the next meeting of the Lord President's Consultative Committee.

The rest of the minutes discussed:

21. Simon Separation Machines (Metropolitan-Vickers).
22. Work at Cambridge Under Dr Feather.
23. Work in Germany.
24. Progress Reports.
25. Visits to U.S.A.
26. Fusing.

Of course, the most famous scientist of the day – ranking alongside Sir Isaac Newton FRS in popularity, Albert Einstein ForMemRS, (see figure 7.4), was working on the theoretical aspects of nuclear fusion.

Even before he was awarded the Physics Nobel Prize (1921), Einstein was arguably the best known scientists – at least to the average person in the street. Indeed US President F D Roosevelt urged him to establish an atomic research project as early as 1939.

Figure 7.4 Albert Einstein.

Later in the year (1942) Sir David Pye FRS wrote to Sir Wallace Akers FRS raising the matter of fusing the atomic bomb – even though it was to be almost three years before the bomb was built. Fuses were a major preoccupation in all armaments as we have already seen.

Ministry of Aircraft Production
Millbank, S.W.1.

13th October 1942.

My dear Akers,
Tube Alloys.

As you know, the bomb scheme is the aspect of the Tube Alloys project in which the Ministry of Aircraft Production is likely to be mainly interested. In this respect I feel that I have the responsibility to the Ministry for ensuring that the many difficult problems involved in fusing a T.A. bomb are fully considered by the appropriate authorities at a sufficiently early stage. I appreciate, of course, that it would serve no useful purpose to bring the experts together until ideas on the characteristics and performance of the bomb are sufficiently crystallised.

The matter is, I think, in the province of the Technical Committee rather than of the Consultative Council and, as the work has now been in progress for almost twelve months under the new arrangements, it seems appropriate to bring it to your notice as Chairman of the Technical Committee.

When it is considered that the time has arrived for the fusing problem to be investigated, I should be glad if this Department could be given the opportunity to take part in any discussions which may arise, or be represented on any Sub-Committee which may be formed to handle the work.

Yours sincerely,
(signed) D.R. Pye FRS
Directory of Scientific Research.

The reply a week later reads:

Directorate of Tube Alloys
17th October 1942.

Dr D.R. Pye FRS,
Ministry of Aircraft Production, D.S.R.
Millbank,
S.W.1.

Secret.

My dear Pye,
Tube Alloys – Fusing of Bomb.
Thank you for your letter of October 13th pointing out that your Ministry has obvious responsibilities in connection with the problem of fusing a T.A. bomb.

I quite agree with this and, when the time comes, we will certainly get into the closest touch with your Department.

I may say here that it is not yet time to do any work on this, because the necessary physical data are not available. There is, however, every indication that the problem will be much less difficult than we had thought, which is all to the good.

I will take this matter up again with you later.

Yours sincerely,
(signed) W.A. Akers FRS[29]

Whilst the effort to produce an atomic bomb gained momentum, some looked to establish medical uses of radiation in treatments for cancer. For example:

British Empire Cancer Campaign

11, Grosvenor Crescent,
Hyde Park Corner,
London, S.W.1.
3rd October, 1944.

To:
Sir Henry H. Dale PRS[30]
President,
The Royal Society,
Burlington House,
London, W.1.

Dear Sir,
National Radiological Institute for Biological and Medical Research.

The Scientific Advisory Committee of the British Empire Cancer Campaign has recently given fresh consideration to the provision of facilities for research into the biological and medical properties of corpuscular and penetrating radiations in relation to the cancer problem. Acting on its advice the Grand Council of the Campaign considers that the time has arrived when concerted action should be taken not only by organisations interested in this particular field of study, but also by those national bodies who are concerned with the wider aspects of biological and medical research.

To this end the Campaign invites the Minister of Health, the Council of the Royal Society, the Medical Research Council and the National Radium Commission to co-operate with it in promoting the foundation of a National Radiological Institute for investigating the biological effects and therapeutic value of (a) extremely penetrating radiation, (b) neutrons, (c) artificially prepared radio-active substances.

In the event of this proposal receiving favourable consideration, the Campaign would like to suggest that the respective parties should each nominate three representatives to form a joint Committee to consider the practicability of the project and to formulate a scheme.

Prior to the outbreak of the present war, the Medical Research Council, the National Radium Commission and this Campaign had set up a joint Advisory Committee to consider the location and installation of a cyclotron to be devoted to the above mentioned objects, the cost of which it was confidently anticipated would be met by a private donor. Although this

Committee held some preliminary meetings its activities were terminated by the outbreak of hostilities and it never reported.

The developments in this field since 1939, the invention and manufacture of practical apparatus capable of producing X rays in the 5 million – 100 million volt region (the betatron), the additional knowledge and experience gained by American workers and the projected organisation of National Cancer Centres in this country, all go to support the view that a larger scheme than that originally contemplated should now be launched. We are accordingly directed to ask for your favourable consideration of this proposal and your co-operation in implementing it.

Yours faithfully,
(Signed)   Hailsham
Chairman of Grand Council
(Signed)   Horder
Chairman of the Scientific Advisory Committee
(Signed)   J.P. Lockhart Mummery,
Honorary Secretary.[31]

As some scientists looked for medical breakthroughs others were transferred to the Canadian Tube Alloys projects. The following, dated 12 April 1945, gives just a glimpse of the high calibre British Chemical Staff, working in Montreal, note that there were also physicists and mathematicians.

Memorandum re: United Kingdom Chemical Staff

At the present time the United Kingdom professional personnel engaged on the chemical aspects of the problems studied by the Akers technical mission comprises:

Dr C.B. Amphlett, Dr W.J. Arrol, Dr K.F. Chackett, Mr G.B. Cook, Mr A.C. English, Dr A.H. Gillieson, Mr B.G. Harvey, Dr H.G. Heal, Dr A.G. Maddock, Mr G.R. Martin, Mr F. Morgan, Dr W.K. Musgrave, Dr C. Reid, Dr H. Seligman, Mr J. Sutton and Sir Geoffrey Wilkinson FRS (Chemistry Nobel Laureate 1973).

In addition to these members Dr L.T. Roberts and Dr R. Wilkinson are engaged on special problems allied to the chemical field. Further Dr B.L. Goldschmidt and Dr J. Gueron, let by the French authorities, are also temporarily available to the United Kingdom organisation. Finally Dr N. Miller and Dr M. Lister of the Canadian staff are comparatively recently from England, and in each case their return has been delayed by the War. They are believed interested in returning to England after the conclusion of the War.

It appears interesting to examine how this staff will fulfil the needs of any future United Kingdom organisation. It is necessary to consider how many of this group expect to become engaged in other work after the War, how completely each section of the work and each specialised technique is being covered by the UK representatives, and finally whether any alterations seem desirable. No consideration has been given to possible personnel outside the Canadian organisation.

Discussion of the future of the various members of the team is necessarily inconclusive, particularly in view of the fact that no knowledge of the conditions or form of a future UK organisation is available. The table given below [as figure 7.5], which must be emphasise, represents only the private opinions of the writers derived from discussions with members of the group, endeavour to give some idea of the future distribution of personnel.

I	II	III	IV
Expect to remain in the UK organisation	Might remain if conditions compare favourably with universities	Expect to leave for a university but available for consultation etc.	Expect to leave for other work
Dr C.B. Amphlett	Dr A.G. Maddock	Dr W.J. Arrol	Dr H.G. Heal
Dr K.F. Chackett	Mr G.R. Martin	Mr B.G. Harvey	Dr W.K. Musgrave
Mr G.B. Cook	Dr C. Reid	Mr G. Wilkinson	Mr J. Sutton
Mr A.C. English			
Dr A.H. Gillieson			
Mr F. Morgan			
Dr H. Seligman			

Figure 7.5 Potential career choices for UK scientists sent to Montreal.

It is certain from this table that Dr Seligman is the only senior chemist who may certainly be expected to be available for a UK organisation. This is largely due to the extreme weakness of the UK team senior chemical personnel. Drs Goldschmidt and Gueron being at present available masks this weakness, and although the writers think Dr Goldschmidt might be persuaded to join the UK organisation for the period of its initial establishment it is probable that he will eventually join Dr Gueron in France, the latter probably returning at the earliest opportunity. It might also be noted that with two exceptions none of the other members of the group have research experience except in the T.A. organisation.

Considering next the various fields of investigation covered these may roughly be divided as follows:

No	Nature of research		Personnel trained
1a	Plutonium studies	Extraction studies	Morgan
1b	Plutonium studies	General chemistry (tracer and micro studies)	Harvey, Heal, Maddock, Morgan
2a	Uranium 233	Extraction researches and purification	English, Maddock, Musgrave, Sutton, Wilkinson
2b	Uranium 233	Extraction plant	Reid
3		Chemical radiation effects	Amphlett
4		General radioactive technique	All members
5		Rare gas technique	Arrol, Chackett
6		Experience at operating pile (projected program)	Amphlett, Cook, Maddock, Morgan
7		Special analytical studies and spectroscopy	Gillieson, Seligman, Sutton

Figure 7.6 Fields of investigation.

Drs Gueron and Goldschmidt have purposely been omitted from this list on account of the temporary nature of their position. In addition to the above, Dr Roberts is working on mass spectrometry with Dr H.G. Thode FRS, and Dr R. Wilkinson is studying the problems of water treatment and control. Further Drs Lister and Miller who might rejoin the UK team are working on sections (2b) and (4) respectively of the above list.

It is concluded from the above list that the UK representation is very weak in the ^{233}U extraction plant group (2b), in the chemical radiation effect group (3) (although this position may be relieved by recovering Drs Lister and Miller) and that not enough senior UK personnel will be getting the necessary operating pile experience to provide the nucleus of the UK chemical team.

On the basis of these observations it is suggested:

1. That one more research chemist, with good record and four or five years research experience be found in the UK and sent to join the team and work at the pile site.

2. That consideration be given to the scheme, already adopted by the Canadians, of selecting one or two established university research

chemists and arranging for them to spend six months or one year's leave of absence from their university in this organisation.

3.   The low representation of the UK group in the extraction plant group is most serious and means of supplementing it should be found.

4.   The urgent need for the early selection of a suitable head of the chemical section of the UK organisation must be reemphasised.[32]

Also in 1945 there was a memorandum outlining the work undertaken in England on the Tube Alloys Program.

Notes on TA Scientific Staff Organisation

This memorandum sets out briefly the arrangements at present in force for recruiting scientific staff, determining their conditions of appointment and allocating them as required to various branches of the project.

Work is at present being carried out in the following places:

I.   Canada – a joint research team of UK based and Canadian scientists is working in Montreal under the auspices of the National Research Council. Professor Sir James Cockroft FRS is the Director (Physics Nobel Prize 1951), and headquarters are at the Montreal Laboratory.

II.   USA – numbers of UK based scientists are working with the Americans at various laboratories and plants in the United States. Professor Sir James Chadwick FRS (Physics Nobel Prize 1935) is British Liaison Offices, and administrative headquarters are at Washington. The organisation is styled the Technical Section of the British Supply Council of North America.

III. United Kingdom
       a)   Research is being carried out at the following universities under contract with the Department of Scientific and Industrial Research
            •   Oxford (Professor Sir Franz Simon FRS).
            •   Cambridge (formally Professor Norman Feather FRS; new arrangements are in hand).
            •   Birmingham (Dr Jones, under Professor Simon: Professor Sir Walter Norman Haworth FRS, Chemistry Nobel Laureate 1937).
            •   Liverpool (Sir Joseph Rotblat FRS Peace Nobel Laureate 1995).
            •   Durham (Dr Eugen Gluckauf FRS).
            •   Manchester (Dr Samuel Tolansky FRS).

The leaders of these research teams are under the direction of the Director of Tube Alloys (Sir Wallace Akers FRS).

b) Research contracts also exist for the conduct of work by the following industrial organisations and research associations
- Metropolitan-Vickers.
- General Electric Company.
- Imperial Chemical Industries.
- Printing and Allied Trades Research Association.
- British Non-Ferrous metals Research Association.[33]

The memorandum is more extensive than that quoted above and goes on to cover issues relating to recruitment, salary etc.

Other Fellows who worked on the Tube Alloys Project include:

- Donald Watts Davies FRS 'was assigned to a group under Sir Rudolf Peierls FRS at Birmingham, working on the atomic bomb. His supervisor was the subsequently notorious Klaus Fuchs. His duties involved him in working at ICI in Billingham, where a ^{235}U separation plant was being developed.'[34]
- Edgar William Richard Steacie FRS 'was appointed Deputy Director, a post he held from 1944–1946, while retaining his duties as Director of the Chemistry Division in Ottowa.'[35]
- Friederich Adolf Paneth FRS 'was Professor of Chemistry and Director of the Laboratories in the University of Durham until 1943 when he was put in charge of the chemistry division of the Joint British-Canadian Atomic Energy team in Montreal.'[36]
- Herbert Wakefield Banks Skinner FRS 'joined a group of British physicists led by Oliphant (in 1943) to whom he acted as Deputy, working at Berkeley California under the late E.O. Lawrence on the electromagnetic separation of the uranium isotopes.'[37]
- Nichol Kurti FRS 'joined a team working on the separation of uranium isotopes by gaseous diffusion.'[38]
- Niels Henrik David Bohr ForMemRS (Physics Nobel Prize 1922). The following is taken from Bohr's Royal Society Biographical Memoir: 'after his arrival in Stockholm arrangements were made by the Director of Tube Alloys and the British Intelligence to transport Bohr and his younger son Aage, also a physicist, to London and on 6th October 1943 he was flown in an unarmed Mosquito back to England being accommodated in the empty bomb rack. He was unconscious for most of the journey to England owing to his not hearing the instructions to put on his oxygen mask. However he recovered consciousness by the time he reached Scotland and was thence flown to London … he was met by Chadwick and other British scientists and learnt about the atomic energy developments and the state of Anglo-American relations … it was arranged after some argument between

Chadwick and General Groves that the Bohrs should visit the United States –Neils as a Consultant to Tube Alloys and his son as a Junior Officer of the DSIR. On 26th August 1944 Bohr met Roosevelt privately after preparing a memorandum which was transmitted to the President by Frankfurter … The essential point of Bohr's presentation was that the Russians should be informed about the existence of the bomb … However the outcome of the Hyde Park meeting between the President and the Prime Minister of September 1944 was very different and the participants decided that the atomic bomb project should continue to have the utmost secrecy.'[39]

There were also Fellows who relocated to the United States to work on the Manhattan Project – the American atomic bomb program, as well as some American scientists who were Foreign Members of the Royal Society. They include:

- Enrico Fermi ForMemRS (Physics Nobel Prize 1938) 'moved to Los Almos to join Oppenheimer on the atomic bomb project.'[40]
- Eugene Paul Wigner ForMemRS (Physics Nobel Prize 1963) 'and his team designed large reactors that could produce practical quantities of polonium.'[41]
- Glenn Theodore Seaborg ForMemRS (Chemistry Nobel Prize 1951) 'headed the Metallurgical Laboratory chemistry group, which was responsible for devising plant processes for chemical purification of plutonium for the World War Two Manhattan Project to develop an atomic bomb.'[42]
- Harold Clayton Urey ForMemRS (Chemistry Nobel Prize 1934) 'was director of the SAM Laboratories (the code name for the Columbia Division of War Research).'[43]
- J. Robert Oppenheimer ForMemRS 'accepted a commission as a Lieutenant-Colonel of the US Army and appointed Director of the Manhattan Project.'[44]
- James Franck ForMemRS (Physics Nobel Prize 1925) 'joined the Metallurgical Project in Chicago, which formed part of the atomic bomb project. He was put in charge of the chemistry division.'[45]
- Maurice Hugh Frederick Wilkins FRS (Physiology or Medicine Nobel Prize 1962) 'vaporised uranium metal with the ultimate aim of fractionating its isotopes and concentrating the more fissionable material. This effort was continued in Berkeley in 1944–45 after he was carried off with the rest of Oliphant's team to California as part of the on-site British contribution to the Manhattan Project.'[46]
- Richard Phillips Feynman ForMemRS (Physics Nobel Prize 1965) 'became a member of the group working with Robert R. Wilson on the electromagnetic separation of uranium-235 and uranium-238 by using the "isotron", a device that the experimentalists at Princeton had developed for accelerating beams of ionized uranium, and trying to separate the isotopes by bunching

them by applying a high-frequency voltage to a set of grids partway down a linear tube.[47]

- Robert Sanderson Mulliken ForMemRS (Chemistry Nobel Prize 1966). 'The headquarters of the Manhattan Project was established at the University of Chicago, with Arthur H Compton as director and Enrico Fermi as leader of the research team. Mulliken was given the job of Director of the Information Division.'[48]

- Sir Mark (Marcus Laurence Elwin) Oliphant FRS, 'worked at the Berkeley Laboratory of the University of California on the electromagnetic method for separating the ^{235}U from natural uranium.'[49]

History records the successful construction and deployment of two nuclear bombs. Figure 7.7 is the dust cloud from the bomb dropped on Nagasaki.

Figure 7.7 The mushroom cloud from the bomb dropped on Nagasaki.

Bouncing Bomb

The attack on the Ruhr valley reservoir dams will forever be associated with Sir Barnes Wallis FRS the scientist, (see figure 7.8), Wing Commander Guy Gibson VC who was in command of 617 squadron raid on the Ruhr dams and the bouncing bomb.

His initial concept, however, was not to produce a bouncing bomb, but to produce a sufficiently large bomb that it would be effective at destroying tanks of fuel stored underground or reservoir dams.

Sir Barnes wrote up his idea in a document titled *A Note on a Method of Attacking the Axis Powers* written in 1940, a copy of which is now stored in the National Archives, the summary of which is reproduced below.

Figure 7.8 Sir Barnes Wallace. (© Godfrey Argent Studio.
Reproduced with permission)

Summary

1) The Bomb Armament of all the belligerent Air Forces in this War consists of relatively small bombs designed to attack surface targets such as factories and houses.

2) This form of attack if effectively countered by Dispersal. It is becoming impossible to destroy simultaneously **all** the factories and **all** the generating stations all over the continent of Europe.

3) All these factories are however dependent upon a relatively few highly localised stores of energy in the form of coal, oil and water power; air attacks on this country are dependent upon large stores of petrol buried in tanks many feet underground.

4) These stores of energy are so concentrated and so massive that they cannot be dispersed, but also they are invulnerable to the present type of bomb armament.

5) This paper shows that:
    a) These stores of energy are vulnerable to very large bombs.
    b) By sterilising their stores of energy the industries of Germany and Italy can be quickly paralysed.
    c) The very large bomb and appropriate aircraft are practicable and can be produced in this country.[50]

Later in the year, Sir Barnes wrote a letter to Air Marshall A.W. Tedder. This must have been considered a breach of protocol given the differences in their respective activities. Dated 4 October 1940, and written on Vickers-Armstrongs Limited notepaper (Sir Barnes worked for them during the War), the letter further expands on the somewhat theoretical paper mentioned above.

Dear Tedder,

This letter is in the nature of an interim report to you on the progress which we have made with the 10-ton bomb.

We have not yet sent in our proposal, as, in view of the large number of objections to a bomb of this size which have been raised by many of the people to whom I have described the idea, it seemed to me essential that I should do enough work on the bomb side of the proposal either (1) to convince myself that I had been mistaken in my original conception, or (2) to show that the idea had sound foundations.

To so this it has been necessary for me to get in touch with a number of people each of whom seem likely to have some specialised knowledge bearing upon the question, and I have now had consultations with the following:

1.   English Steel Corporation at Sheffield on the subject of the manufacture of the bomb casing and question of the steel to be used. On this visit I took with me Wing Commander Sealy in case any questions connected with the filling of the bomb should arise and his presence proved very useful.
2.   Professor G.I. Taylor of Cambridge to discuss the law connecting the charge weight of a bomb and the destructive work which it is capable of doing.
3.   Professor J.D. Bernal FRS, Ministry of Home Security, Prices Risborough, in connection with the destructive power of bombs, both in earth and water.
4.   Professor G.T.R. Hill at Exeter to discuss points in connection with the structural design and stressing cases for the bomb casing.
5.   Sundry visits to and from Vaughan Williams and his assistants at Thames House in connection with the question of probability and ballistics.
6.   Mr E.F. Relf FRS of the National Physical Laboratory in connection with the drag coefficient of bombs travelling at supersonic speeds.

The investigations which I have been making fall under three headings:
i.   The design and manufacture and filling of the bomb casing.
ii.   The destructive power of bombs in general and with particular reference to special types of target.
iii. The ballistics of high altitude bombing.

i.   English Steel Corporation are able to supply a steel casing which will give between two and three times the strength of the material which is at present used for the manufacture of bomb cases. The steel they propose had a normal ultimate tensile strength of 60–70tons/sq.in. but can actually be made to an ultimate tensile strength of 90tons/sq.in. as compared with 35tons at present, and the officials at the English Steel Corporation do not anticipate any abnormal difficulties in the supply or handing of this material. Wing Commander Sealy did not seem to fear any abnormal difficulties in filling and it is easy to prepare the interior by grinding local rough places followed by varnishing if a smooth surface is required.

ii.   We have acquired a considerable amount of information on the destructive power of large charges of high explosive and special examples have been worked out showing what the large bomb will do. On this point I have found much useful information in the library of the Institution of Civil Engineers where handbooks on mining are available, describing the effects of charge weights up to as much as 100,000lbs in a single charge. Having regard to the targets which I had specially in mind when making my original suggestion, I have obtained drawings from firms specialising in the design and construction of large dock gates and hydro-electric installations and actual examples are now being worked on to show the minimum size of bomb which is required to produce destructive damage to targets such as these.

iii. I have made an independent investigation into the ballistics of large mass bombs released from 40,000ft and find that subject to the actual drag coefficient being substantially that which has been assumed in our calculations the large bomb will certainly attain supersonic speeds at impact. This point I think is one of great importance, as the target which first drew my attention to the necessity for a large bomb; that is the deeply buried reinforced concrete oil storage tank, can only be reached by a bomb having sufficient kinetic energy to enable it to penetrate 60–70ft below the surface of the ground. I was told by 1936 by an old school fellow who was Chief Experimental Chemist to the Shell-Mex Group at their research laboratory in California, that his company had just finished delivering sufficient petrol for two years full War consumption in Germany, and I believe that it will prove quite impossible to reach such tanks by any other means.

The equations of motion used by the Air Ministry for calculating their ballistic tables are arranged to eliminate the mass of the bomb; obviously a desirable thing when bombs of all sizes have to be dropped. We are able to show that, whereas for example a 500lb bomb released at 40,000ft is unable to break through the sonic barrier and is actually going slower at ground level than it is at 10,000ft, the 10-ton bomb has a sufficiently high cross-sectional density to pass through the barrier and is still accelerating at the time of impact which should take place at a velocity of approximately 1,400ft/sec. It will clearly be necessary to introduce a correction for the mass of bomb when altitudes greater than 10,000ft are considered.

In order to confirm the drag coefficient used in our calculations we are making a large scale model of the bomb, the design of which has now taken definite shape and we have written to the Director of the National Physical Laboratory, on Relf's advice, asking if they will be good enough to test a model of the bomb in a supersonic jet. A combination of the results from these two models should give a drag curve for the full range of speeds which the bomb will pass through.

I think I am now sufficiently far advanced to be able to say that while some of the work which we have undertaken is difficult and must obviously be subject to empirical corrections, all the evidence goes to confirm my original conclusion that only a very large bomb can do effective work against such targets.

I was disappointed to learn from Professor Hill that he had had a note from the Air Ministry saying the questions I had discussed with him were outside the scope of his department. As a result of our meeting I had formed the opinion that his assistance would prove most helpful. The design of a bomb casing this size is a matter for a structural engineer rather than a bomb expert. To determine the bursting effect at impact Professor Hill had suggested the use of models with some approach to dynamic stability to the full scale by the use of soils varying from mud upwards, all of which were available on the ranges at Exeter, and I should be grateful if we might continue to consult Professor Hill when necessary and to know whether in view of the difficulty that we shall experience in attaining dynamic similarity in small scale models the proposal to use the Exeter ranges with their wide variation of soils might not be reconsidered.

Yours sincerely,
B.N. Wallis[51]

The drawback to this approach was the need for very large wingspan aircraft, additionally the drag on them from the bomb was not insignificant. A Committee was established to evaluate this strategy. The first two items of the meeting held on 10 December 1941 are interesting:

AAD Committee

Minutes of the 3rd Meeting held at the Ministry of Aircraft Production on 10th December 1941.

Present:
Mr R.S. Capon   Chairman
Professor Sir Geoffrey Taylor FRS
Professor W.N. Thomas
Sir William Glanville FRS
Dr A.H. Davis
Sir Barnes Wallis FRS
Mr W.H. Stephens   Secretary

In the unavoidable absence of Dr Pye FRS, the chair was taken by Mr Capon.

1. The Chairman recalled the suggestion at the last meeting that it might be more economical to rely on direct hits using a large number of small bombs rather than one large bomb, in view of the very slight gain in

effective width of the target (measured by the horizontal projection of the sloping face) for a large increase in size of bomb, and said that a series of experiments had subsequently been made with contact charges at the Road Research Laboratory. Dr Davis then described the experiments in which 2oz charges of PAG had been exploded in contact with and at short distance from lightly reinforced concrete slabs. It had been decided to use flat slabs because model arches thick enough to withstand contact explosions assumed that for contact charges the effect was independent of the radius of curvature. The results indicated an appreciable difference between slabs and arches for charges at small distances, a bigger charge being required for the latter, and showed that water tamping was more effective than earth tamping. Photographs of the damage indicated a different type of failure for arches and slabs.

2.   Considerable discussion followed on the validity of the experiments, during which Mr Wallis pointed out that the charges had not been buried below the camouflet depth in earth and might therefore lose some of their effectiveness, whereas in the case of water tamping there was probably an appreciable gain due to reflection from the back face of the water cavity, which was only about 1½ft from the slab.[52]

Clearly more realistic experiments were needed.

Sir William Glanville FRS, one of the attendees of the above meeting developed the use of models for the study of a wide variety of War problems, playing a leading part in the research resulting in the breaching of the Mohne and Eder dams, in the development of concrete piercing bombs, the invention of plastic armour, the measurement of blast pressure from bombs, the movement of torpedoes and the demolition of structures. The above appears in the Royal Society Fellows database.[53] A site was chosen in Wales, as documented in a letter from Sir Barnes to W.H. Stephens of the Ministry of Aircraft Production.

21st July 1942
Dear Mr Stephens,

Further to my note written on your receipt form of 16th inst. I will attempt to reach Nant-y-Gro on Friday if possible. At present I am working under heavy pressure that it seems unlikely that I shall be able to get away, but I shall be glad to be kept in touch with the results of the experiments and to see any films or other photographs which may be obtained.

May I point out that the notice which you give is very short and that the Committee does not seem to have been consulted either in planning the

latest series of experiments or in deciding upon the final treatment of the
Nant-y-Gro Dam.

Yours sincerely
Barnes Wallis[54]

More is known of these experiments now (April 2013) than may have been
known for some time as some of the experiments were photographed, but the
photographs were subsequently lost. They came to light earlier on this year in
the National Archives where they have now been catalogued.

These experiments are in line with a change in thinking regarding bomb
deployment, as documented below.

Wallis Spherical Bomb

1. In the middle of 1942 Mr Wallis of Vickers, Weybridge, evolved the
    idea of a spherical bomb which would be dropped from aircraft on to
    water, and would bounce along the surface in much the same manner
    as 'ducks and drakes' thrown on to still water.

2. The principle was first proved in tank experiments at the NPL and
    these looked so promising that we proceeded to full-scale experiments
    on a Wellington aircraft. These have now reached a stage at which
    we can say that the principle is proved. Although the full range has
    not been obtained, the bomb has been made to bounce and to cover
    distances of well over 1,000 yds. The experiments have been carried
    out at Chesil Beach with the necessary assistance of the Admiralty
    Research Department, who are, therefore, in full possession of the
    experimental data.

3. We are now emerging from the experimental stage into the develop-
    ment phase and for this purpose an order has now been placed for 250
    bombs to be carried in Mosquito aircraft. The attached short note from
    DSR gives you more particulars.

4. The object of this note is to make you acquainted with what is going
    forward. There are other applications of this bomb but they are being
    dealt with as an entirely separate matter.

Signed
CRD
4th Feb 1943[55]

These experiments were conducted with a range of spheres of different
surface characteristics. Figure 7.9 is of the surviving spheres kept on display at
the RAF Museum in Hendon.

Figure 7.9 Some of the practice spheres. (Author's photograph)

Of course the final bomb shape was changed to be cylindrical and not spherical. Fortunately, some of the practice bombs, which were filled with concrete, survived the experiments. One is kept in the Derwent Reservoir, the site of some of the practice bombing runs, (see figure 7.10), and one in the RAF Museum at Duxford, see figure 7.11.

Incidentally, the RAF Museum at Duxford display the bomb under a bomber, which provides a sense of scale.

Figure 7.10 Derwent Reservoir in Derbyshire. (Author's photograph)

Figure 7.11 One of the remaining practice bombs. (Author's photograph)

History records that the bombs and aircraft modifications were completed in time and that the raid was successful. It took place on 16–17 May 1943. The order was given for the raid on the 15 May 1943 from the Air Ministry Whitehall AX457 to Bomber Command which read:

Most Secret

Operation CHASTISE immediate attack of targets XYZ approved. Execute at first suitable opportunity.

Most Immediate[56]

The briefing papers (without the appendices) for the raid had been reproduced as Appendix X for completeness.

Gas/Biological warfare

The threat of gas warfare by the German troops was sufficient for the Government to continue research and development. Accordingly there were centres of gas warfare research, some of the Fellows from which are listed below:

Name	Activity
Addison (Cyril Clifford)	Worked on vesicant gas production.[57]
Bailey (Kenneth)	Joined Dr Malcolm Dixon's team (early in 1943) working on vesicant (tear gas) action under the Ministry of Supply and on the interaction of mustard gas with the yeast hexokinase.[58]
Cameron (Sir Gordon Roy)	Seconded to the Chemical Defence Experimental Station at Porton and moved to the Biological Warfare section under Fildes in April 1943.[59]
Evans (Sir Charles Arthur Lovatt)	Worked at the Chemical Defence Experimental Establishment at Porton Down (1939–1944).[60]
Everett (Douglas Hugh)	Worked with Sir Cyril Hinshelwood FRS on respirator charcoal.[61]
Frank (Sir Charles)	Gas mask smoke filters and smoke screens (awarded an OBE in 1946 for his war-work).[62]
Gibson (Charles Stanley)	Became Senior Gas Advisor for the London Region and subsequently for Region 12.[63]
Glenny (Alexander Thomas)	Involved in the provision of vast amounts of tetanus and gas-gangrene antitoxins.[64]
Henderson (David Willis Wilson)	Became a member of staff of the Bacteriology Department of the Lister Institute (Sept. 1939). Seconded for special duties with the Ministry of Supply and found himself working either at the Chemical Defence Experimental Establishment, Porton Down or at the Lister Institute at Elstree. Henderson crossed the Atlantic in 1943 to assist the Americans with the culture of virulent organisms and their aerosol testing. For the remainder of the War he travelled back and forth to America. He was awarded the US Medal of Freedom, Bronze Palm in 1946.[65]
Jones (Sir Ewart Ray Herbert)	Undertook secret War research, notably for the Ministry of Supply on the preparation of lachrymators.[66]
Spring (Frank Stuart)	Protection of the civilian population against gas attacks.[67]
Wormall (Arthur)	Together with G.E. Francis, J.C. Boursnell, T.E. Banks and F. Ll. Hopwood he investigated the reactions of mustard gas with proteins using radioactive reagents.[68]

Figure 7.12 Some of the Fellows working on gas warfare problems.

8   Engineering, transport improvements and raw materials

Understandably, improvements were not confined to armaments. Research and development was undertaken in all of the Services for all types of equipment, including uniforms, some of which have already been mentioned. Perhaps such a broad subject as engineering improvements needs a focussed approach in order to appreciate the broad development work undertaken in just one branch of the Services, on the understanding that equal work was in hand elsewhere.

Arguably, the speed of aircraft development, or the introduction of new aircraft types, was greater than the equivalent transport development in the other branches of the Services. Additionally, the numbers of aircraft and the time interval between airfield and enemy action was shorter, perhaps resulting in many more opportunities to further develop them.

Aviation improvements

By way of illustration the work of the Air Fighting Development Unit (AFDU) from March 1940 to September 1942 best illustrates the rate of change in aircraft modification and new introductions, see figures 8.1 and 8.2.

Fighters	Bombers	Other types
12-gun Hurricane II	Boeing B-17C	Glider (day and night)
Airacobra	Boston D.B.7	Stinson Vigilant (Army co-op)
Defiant II	Boston III	
Havoc 1 night fighter	Fortress B-17F	
Lockheed P-38F	Glenn Martin Maryland	
Mohawk	Halifax	
Mustang	Lancaster	
Spitfire IX	Manchester	
Spitfire VI (pressure cabin)	Mitchell B-25	

Fighters	Bombers	Other types
Tomahawk	Mosquito	
Typhoon	Stirling	

Figure 8.1 Some of the tactical trials for new aircraft. (National Archives AIR 2/1933)

Captured Enemy types
Fiat C.R.42
Focke Wulf 190
Heinkel III
Junkers 88
Messerschmitt 109E
Messerschmitt 109F
Messerschmitt 110

Figure 8.2 Some of the tactical trials for captured enemy types. (National Archives AIR 2/1933)

The miscellaneous tactical trial included:

- Airacobra – boost control (Claudel Hobson).
- American pairs.
- Harness for 'K' type dinghy pack.
- Hurricane – attack on formations with bombs.
- Hurricane and Spitfire – camouflage.
- Kittyhawk – Claudel Hobson boost limitator.
- Mustang – Claudel Hobson boost limitator.
- Mustang – Deloo Remy automatic boost.
- Mustang – new sliding hood.
- Mustang – periscopic mirror and modified gun sight.
- Mustang – periscopic sight.
- Mustang – rear view mirror.
- New type of 'G' mask and 'C' helmet.
- New type of goggles.
- Perspex and Mk VII goggles.
- Safety markings of flying and engine instructions.
- Spitfire – D.H. hydromatic propeller.
- Spitfire – increase in engine power at high altitudes by the use of liquid oxygen.
- Spitfire – increased boost for Merlin 45 (+16).

- Spitfire – inertia device.
- Spitfire – interconnected throttle propeller and mixture controls.
- Spitfire – IX comparative trials .447 and .42 reduction gear.
- Spitfire – new type hood.
- Spitfire – production periscopic sight.
- Spitfire – R.A.E. anti-negative 'G' carburettor.
- Spitfire – R.A.E. improved inertia device.
- Spitfire – rear vision mirror.
- Spitfire – Rolls Royce anti negative 'G' carburettor.
- Spitfire – tactical handling in relation to all up weight.
- Spitfire – Westland elevator.
- Spitfire and Hurricane – device to prevent Merlin engine cutting during application of negative 'G'.
- Spitfire V – high altitude interception.
- Spitfire VB – comparative trials with wing tips removed and a small modification to the wings.
- Spitfire VC – universal wing.
- Spitfire VI – double hood.
- Spitfire – G.M.2 sight with modified lamp holder.
- Use of flares in defence of bomber aircraft.
- Wellington – protection of bombers against fighter attacks by means of oily smoke.

There were also extensive armament trials for:

- Airacobra 37mm gun.
- Beaufighter cannon.
- Beaufighter modified G.J.3 sight.
- Defiant gyro sight fitted to turret.
- Gyro sight Wellington rear turret.
- Gyro sight Whirlwind.
- Harmonisation stand.
- Hurricane 20mm tracer.
- Hurricane gyro sight.
- Hurricane gyro sight bracket.
- Lockheed P-38F – armament trials.
- Spitfire cannon.
- Spitfire modified G.M.2 sight.
- Spitfire periscopic sighting device.
- Tomahawk.
- Typhoon gun heating for A.G.M.E.
- Typhoon modified G.M.2 sight.
- Typhoon modified link chutes for .303.

- Typhoon periscopic sight alteration of present bracket.
- Wellington III gyro sight in F.N.20 turret.

Other issues surfaced from time to time, including:

To: AFDU Duxford
From: Fighter Command

20mm Tracer Ammunition
It is desired to test the effectiveness of 20mm tracer ammunition for day use. It is therefore requested that AFDU carry out trials with this ammunition and submit a report to this Headquarters in due course.

The report should cover the following points:
i.   Is the ammunition considered to be of use as an improvement to sighting?
ii.   Is the trace of the correct brilliance and duration? (The ammunition is reported to have a very long trace)?
iii. What proportion of tracer ammunition to invisible types is recommended and is this proportion in one gun only or in all guns?
Arrangements have been made for 500 rounds to be despatched to RAF Station, Duxford

Signed for
Air Marshall
Air Officer Commanding-in-Chief
Fighter Command[1]

One might imagine, given the signatory, that this received a high priority.

The Air Fighting Committee met regularly throughout the War. Part of their agenda was to review progress of the Air Fighting Development Unit. This Committee comprised mainly of Air Commodores, Group Captains, Wing Commanders and Squadron Leaders from the Air Ministry. Sir Henry Tizard FRS was also included in their representatives. The Commands included Air Chief Marshall Sir Hugh Dowding, Air Vice-Marshall N.D.K. MacEwen, Air Vice-Marshall T.L. Leigh Mallory as well as officers of the ranks already mentioned. The Establishments were represented by Professor B. Melvill Jones FRS, Dr L.B.C. Cunningham and Mr L.G. Savage.

Committee minutes from the 19th meeting held at the Air Ministry on Monday 12 February 1940 contain the following:

The Chairman (D.W.T.T.) said that the purpose of the meeting was to review the investigations which had recently been in progress at the Air Fighting Development Unit (some of which were not completed), and to discuss the recommendations which had been received from the Commands for the future program; placing the items in order of priority. He mentioned that A.C.A.S. was unavoidably prevented from taking the Chair.

The minutes for this particular meeting are 19 pages long providing detailed discussion of each of the trials undertaken by the AFDU. Little would have been known of this Committee during the War. Nevertheless is demonstrates the commitment at the senior levels of the RAF to ensure state of the art equipment and safety standards.

In some cases the AFDU completed their investigations of German planes and then passed them on to other units for a more detailed analysis of a specific technology. For example, the following is the first part of a summary for a report undertaken at the Royal Aircraft Establishment, Farnborough:

Enemy aircraft Messerschmitt ME109 number 1304
General report on Radio Equipment type FUG7

Summary

This equipment consists of transmitter, receiver, motor generator, combined junction box and power pack, heater resistances, main switch, transmit-receive switch and aerial current meter.

The frequency range covered is from 2.5 to 3.75 megacycles (80 to 120 meters). Telephony only is used. The transmitter consists of a master oscillator, modulator and two output valves in parallel. Grid modulation is applied to the output stage.

The receiver is of the super-heterodyne type consisting of five valves – H.F. amplifier, tetrode frequency changer, two I.F. amplifiers, demodulator by metal rectifiers which also provide AVC and output. All receiver valves are the same type of tetrode – Telefunken RENS 1264. Power.[2]

The same National Archive file contains a report of a Junkers Ju.87.B2. Number S2LM. In this case the aircraft was examined at Ventnor on the Isle of Wight on 9 August 1940. This report also concerns the radio, which in this case was a type FuG VIIa.

Some of the technology implemented by the Air Fighting Development Unit was developed by The Aeronautical Research Committee, which had the following terms of reference:

1) To advise the Secretary of State on scientific problems relating to aeronautics.

2) To make from time to time recommendations to the Air Council as to any researches which the Committee consider it desirable to initiate, and as to any matters referred to them by the Council.

3) To supervise the aeronautical researches at the National Physical Laboratory initiated by them and if requested to do so by the Air Council any other researches connected with aeronautics.

4) To make an annual report to the Air Council of the research work which the Committee consider should be undertaken at the National Physical Laboratory, or elsewhere, together with an estimate of expenditure at the National Physical Laboratory.

5) To investigate the causes of such accidents as may be referred to them by the Air Council and to make recommendations as to the prevention of accidents in the future.

6) To promote education in aeronautics by co-operating with the Governors of the Imperial College and in any other way within their power.

7) To assist with advice any research carried out by or on behalf of the Aeronautical Industry and to make available any information of value to the Industry so far as is compatible with public interest.

8) To make an annual report to the Secretary of State for Air.[3]

Membership of the Committee in 1939 was:

Independent

Sir Henry Tizard FRS (Chairman), Sir Bennett Jones FRS, Lord Blackett FRS, Professor W.J. Duncan FRS, Sir Ralph Fowler FRS, Professor G.I. Taylor FRS, Sir George Thomson FRS.

Official

Dr E.V. Appleton FRS (DSIR), Sir Charles Darwin FRS (NPL), Dr H.J. Gough FRS (War Office), Mr A.H. Hall (RAE), Mr N.K. Johnson (DMO, Air Ministry), Sir David Pye FRS (DSR, Air Ministry), Mr C.S. Wright (Admiralty).

The Aeronautical Research Committee had several Sub-Committees and Panels, a full list of members and terms of reference for which appear as Appendix XI. Briefly they were (in 1940):

Sub-Committees (with Panels where appropriate):

1) Aerodynamics.
   i)   Airscrew Panel.
   ii) Fluid Motion Panel.
   iii) Kite.
2) Alloys.
3) Elasticity and Fatigue.
4) Engine
   i)   Lubrication Panel.
5) Fleet Air Arm Research.
6) Metrology.
7) Oscillation.
8) Plastics.
9) Seaplane.
10) Stability and Control.
11) Structure.

There were also three Panels associated directly with the Aeronautical Research Committee. They were:

- Free Flight.
- Navigation.
- RAE High Speed Wind Tunnel.

The titles provide a fair indication of the work of this Committee. Add to that the high calibre of the scientists and engineers and one can be assured that the Aeronautical Research Committee provided far reaching support across all aspects of aeronautics. It is worth looking at some of this research in more detail as aspects of aeronautical research were new in World War Two. For example aircraft carriers in the Fleet Air Arm were used for the first time in warfare. Unique research opportunities concerning the whole issue of landing on a moving platform required extensive research. Not least the issue of determining the exact relative position of aircraft and landing platform.

Indeed the specific requirements of the Fleet Air Arm were communicated by the DSR of the Admiralty. Titled *Some Problems of the Fleet Air Arm* the report contains the following:

1.0 Role of the Fleet Air Arm

The Fleet Air Arm exists solely as part of the Navy – it has no separate justification; it has no duties independent of the Navy. Its resemblances to the RAF are more obvious than its differences, but it is the latter which cause the Navy most concern since these have necessarily received least attention in the past, and the necessity is increasingly felt for aircraft designed ab initio to meet these needs, dominated as they are by limitations peculiar to aircraft operation at sea.

The Navy is an all-weather service, and therefore desires all-weather aircraft. That is why it dislikes float-planes. It is also the main reason for the existence of the Carrier, since it has been laid down as a prime necessity that the Fleet Air Arm should be 'able to operate over the widest possible range of weather conditions'. It is further required that all essential services must be provided by carriers, since catapult ships are handicapped by their inability to retrieve their aircraft except in reasonable weather. Present policy requires as seaworthy an aircraft as possible, either amphibian or floatplane, in all capital ships and large cruisers. The aircraft in such ships are required primarily for spotting when with the Fleet, and for various duties when operating without carriers.

Later in the report there is a list of requirements:

the characteristics which an aircraft should possess in order to meet the needs enumerated in 2.1 are set below.
1) Non-attacking
    i)   Reconnaissance – to find the enemy.
    ii) Observation – to report his movements.
    iii) Spotting – to direct gunfire (one aircraft for each firing ship).
    iv) Night shadowing – to prevent escape during the night.
2) Attacking
    i)   Torpedo – the principal weapon.
    ii) Dive bombing.
3) Miscellaneous
    i)   Fighters with fixed guns or turret guns.
    ii) A/S. Patrol ahead of the Fleet.[4]

These specific requirements were investigated by members of the Fleet Air Arm Sub-Committee. Unsurprisingly, titles of some of their reports are consistent with the above criteria.

- Autogiro trials (19 January 1940).
- Note on Present Position of Rotating Wing Aircraft Development (24 January 1940).

- Comparison of 'Walrus' and 'Sea Otter' (12 February).
- Note on Deck Landing – methods on reducing the landing run (re-issued February 1940).
- Monoplanes and Biplanes to specification S.24/37 (5 March 1940).
- Airflow over the flight deck of a model aircraft carrier. Dispersal of cliff eddies by forced draught (11 March 1940).
- Limitations imposed by aircraft carriers (31 October 1940).[5]

Other Aeronautical Research Committee's work addressed some of the issues relating to supersonic flight. Written by E.F. Relf FRS, a report was published through the Aerodynamics Sub-Committee of the ARC on 28 May 1943. Whilst most of the report is highly technical, the opening remarks set the scene:

> It is evident that conditions in supersonic flight are different from those to which we are accustomed in dealing with the performance of ordinary aeroplanes, and the present brief report seeks to discover broadly that those differences are, so that a clear picture of the supersonic conditions can be formed.
>
> It is probable that some form of power unit will ultimately be developed with sufficient thrust, at any rate for short exploratory flights at these speeds, but neither the power problem, nor the even more difficult one of control through the sonic speed is here considered. The object is simply to establish the broad nature of the conditions of horizontal flight when the speed is well above that of sound.[6]

At the time of his election to Fellowship in the Royal Society (May 1936), Ernest Frederick Relf was Superintendent of the Aerodynamics Department in the National Physical Laboratory. It is doubtful that as a member of the NPL he would have heard of a young scientist called Frank Whittle who was to change the course of aviation history by inventing the jet engine.

Whittle joined the RAF as a boy apprentice, qualifying as a pilot at the RAF College in Cranwell. He served as a test pilot in addition to a variety of postings within the RAF. His early ideas on jet propulsion were formulated when he was in his early 20s in 1928. There followed a series of jet related milestones:

- Registered a patent for use of gas turbine for jet propulsion (16 January 1930).
- Turbo jet patent granted and published in many countries (1932).
- Formed Power Jets Limited with R.D. Williams and J.C.B. Tinling with the support of O.T. Falk and Partners and B.T.H. Company, Rugby (March 1936).

- RAF Special duty list, attached to Power Jets Ltd as Honorary Chief Engineer (1937–1946).
- Gloster Whittle E 28/39 maiden flight at Cranwell (15 May 1941)
- Power Jets Ltd nationalised (1944).

Initially ridiculed by the Air Ministry as impractical, a jet engine of his invention was fitted to a specially built Gloster E.28/39 airframe, the maiden flight of which took place on 15 May 1941. By 1944 Gloster Meteor jet aircraft were in service with the RAF, intercepting German V-1 rockets. In the same year Power Jets Ltd was nationalised.

There followed a long struggle before Whittle's achievements were officially recognised.

> Request by the Ministry of Supply to the Royal Commission on Awards to Inventors on the question of an award to Air Commodore Frank Whittle
>
> The Ministry of Supply has approached the Commission with the request that they would assess the award to be paid to Air Commodore Whittle in respect of his invention and development of jet propulsion engines for aircraft and gas turbines generally. The Commission have consented to assist the Department in this matter. The members of the Commission have visited NGTE at Whetstone to acquaint themselves with the technical aspects of the case and with the part taken by Air Commodore Whittle in the initiation and development of the internal combustion turbine and, it is presumed, that it is not necessary for a further statement on these matters to be made. Other relevant matters are being placed before the Commission at this hearing, Air Commodore Whittle has agreed to the procedure adopted in this case by MoS.
>
> The Department's interest derives from the Ministry of Aircraft Production and through that Department from the Air Ministry. Air Commodore Whittle has been at all relevant times and still is an officer of the Royal Air Force. The assessment by the Commission of the award is this on behalf of the Air Ministry as well as the MoS. Payment of the award of the Commission will be made from the MoS votes arising in connection with the technical services rendered by that Department to the Air Ministry. Treasury consent has been obtained by MoS to the assessment of the award by the Commission.
>
> Technically, Air Commodore Whittle is not a claimant for the award and the Department's request to the Commission primarily derives from a clause in an agreement, a copy of which has been furnished to the Commission, for the acquisition in April 1944 by the Minister of Aircraft Production of Air Commodore Whittle's interest in Power Jets Limited.[7]

Whittle retired from the RAF in 1948 with the rank of Air Commodore, and that same year he was knighted. The British Government eventually atoned for their earlier neglect by granting him a tax-free gift of £100,000. He was awarded the CBE (1944), CB (1947), a KBE (1948) and OM (1986). The Royal Society elected him a Fellow on 20 March 1947. Four other Fellows spring to mind who were involved with jet engines.

- Alan Arnold Griffith FRS 'joined Rolls Royce devoting the first six months to blade design calculations, performance estimates and a general design study of several alternative turbine proposals for aircraft propulsion. He wrote a paper which became the basis of the Rolls-Royce 'Avon' engines.'[8]
- Geoffrey Bertram Robert Feilden FRS 'joined Power Jets Limited where he managed the test program for the Whittle engine.'[9]
- Hayne Constant FRS 'was Head of Engine Department Royal Aircraft Establishment Farnborough. Constant and Sir Frank Whittle enjoyed a close collaboration on the development of the jet engine.'[10]
- Herbert Brian Squire FRS 'joined the Aerodynamics Department of the Royal Aircraft Establishment Farnborough (early 1934), moving to the Wind Tunnel Section of the Aerodynamics Department to take charge of a newly formed group engaged in the research involved in some promising new developments such as tailless aircraft and jet aircraft.'[11]

Air Commodore Sir Frank Whittle OM FRS was not the only inventor caught up in the ownership of intellectual property during the War. The following was the position for radio and radar inventions.

Preliminary Report on Radio and Radar patented and patentable inventions owned by H.M. Government

Introduction

1. This report which relates to conditions obtaining on 30th November 1945 is concerned with the work already done by the Patents Division of the three Service Ministries, Admiralty, Ministry of Aircraft Production and Ministry of Supply in filing patent application in the radio and radar field, and their future program, with an estimate of their requirements as to personnel.
2. A list of all patent application filed during and since the War, and numbering 389, forms an appendix to this report.
3. The future program includes the following items:
   i.   Preparation and filing of patent applications in the United States of America based on existing UK applications.

ii.   Production of evidence as to the date of disclosure of the inventions concerned under the heading A, to the US Government, in order to claim a date under the proposed US Act.

iii. Preparation and filing of patent application in Canada, Australia, New Zealand, South Africa and India based on existing UK applications.

iv.   Preparation of complete specifications for those UK applications which were accompanied by a provisional specification. This work must be done before US and other overseas applications can be filed.

v.   Prosecution of UK cases already completed.

vi.   Preparation and filing of UK patent applications on important inventions already investigated.

vii   Investigations of new inventions and filing corresponding UK patent applications in suitable cases.[12]

Figure 8.3 outlines the extent of the problem.

Year	Admiralty	Ministry of Aircraft Production	Ministry of Supply	Totals
1940	2	0	0	2
1941	12	0	–	12
1942	14	5	9	28
1943	14	12	8	34
1944	39	25	18	82
1945	50	100	81	231
Totals	131	142	116	389

Figure 8.3 Patent applications filed.

Whilst the jet engine was being developed there was an ever greater need for faster aircraft. One individual in particular decided to investigate how to maximise engine performance through the use of a better fuel. The Royal Society Biographical Memoir for Sir Ronald Holroyd FRS offers the following:

From 1937 he was Research Manager of the Oil Division of Billingham (an ICI plant). When the VI rockets (doodlebugs) were launched against Britain the fighter aircraft of the period were just not fast enough to catch them and shoot them down. Holroyd started an investigation on special fuels – economics not mattering – to boost the power output and from this concluded that butyl benzene was best. Syntheses was determined (ensuring sufficient quantities of raw materials) and one whole reactor system at Billingham was converted to this product, manufactured under the code name of 'Victane'. This boosted the top speed of fighter aircraft by as much as ten mph and enabled many VI's to be shot down over open country before reaching London.[13]

Sir Harold Brewer Hartley FRS also has a relevant entry in his Royal Society Biographical Memoir:

> Appointed Chairman of the Government factory for producing 100-octane aircraft fuel (in 1939). Also a member of the Sub-Committee of the Advisory Council on Scientific Research. As Chairman of the Sub-Committee on Axis Oil he was responsible for advising on oil targets for bombing raids. Other problems which he took close interest in were that of limiting smoke emission by coal-fired merchant ships in convoy, in the development of effective flame thrower fuels, in devices for camouflage by smoke and of water surfaces by coal dust and in the FIDO project for clearing runways of fog.[14]

Whilst not involved in aircraft production, Sir Alfred Grenville Pugsley FRS, nevertheless played a vital role in their production. Pugsley was appointed Principal Technical Officer of the Airworthiness Department (1938) and was later made Head of the Department.[15]

Raw materials

All raw materials were at a premium during the War, it was merely a matter of where they were and how much was needed. Those Fellows with geology backgrounds often found themselves surveying various locations searching out previously unknown deposits of various minerals or re-assessing the commercial potential of abandoned mines. Some investigated synthetic alternatives. For example:

a. Darashaw Nosherwan Wadia FRS 'was appointed to the post of Government Mineralogist in the Government of Ceylon. Produced a succession of reports on particular minerals of significance to Celon's economy.'[16]

b. George Martin Lees FRS 'searched for oil in Britain.'[17]

c. Peter Joseph Wilhelm Debye ForMemRS 'led to an intensive program of studies for rubber substitutes.'[18]

d. Robert Millner Shackleton FRS 'was a geologist in the Mining and Geological Department of the Kenyan government and mapped the Nyeri, South Nyanza and Maralal. During the War he was charged and deeply gored in the thigh by a maddened rhinoceros, fortunately he survived.'[19]

e. Sir (Cyril) James Stubblefield FRS 'undertook detailed re-mapping of coalfields in Britain.'[20]

f. Sir Kingsley Charles Dunham FRS 'worked in the Pennines and further afield, assessing reserves of zinc ore, fluorspar and barytes. He recommended the recovery of zinc from old mine dumps at Nenthead, Cumbria, and a successful mill processing 1000 tons per day was set up here.'[21]

g. The Svedberg ForMemRS 'worked on the development of a Swedish production of synthetic rubber (polychloroprene).'[22]

h. Vincent Charles Illing FRS 'searched for oil, developing exploration and production in Trinidad and Venezuela.'[23]

The National Archives also stores information on other industries whose products were in the national interest. For example:

Committee of Imperial Defence Flax
(previous C.I.D. Paper no D.P.R. 274)
Note by the Chancellor of the Duchy of Lancaster

I attach a copy of a memorandum by the Chairman of the Supply Board making certain recommendations as to flax. Since there is a definite defence interest in this commodity, I have thought it advisable, in the first instance, to submit the paper for consideration of the Committee of Imperial Defence. The matter, however, also raises issues outside the defence sphere and a later reference to the Cabinet would seem to be requisite

Initialled W.S.M.[24]

This note accompanies the aforementioned paper, following which there was a meeting to work through the fine detail.

Norfolk Flax Experimental Station
Transfer to Government Control

Present: the Forth Sea Lord, Mr R. Wemyss Honeyman – Managing Director of Norfolk Flax Ltd (formerly referred to as Mr A), Mr R.D.

Fennelly – Assistant Secretary, Board of Trade (representing the Ministry of Supply), Mr B. Page – Deputy Director of Stores (Admiralty Representative on the Flax Sub-Committee)

he recapitulated the decision of the Committee of Imperial Defence to take over the Establishment as a Government concern, and place it under the control of a single Government Department for future operation, and pointed out that, while the Department responsible in the future would be the Ministry of Supply, it had been arranged that, until the Ministry had been set up and had got into its stride, the Admiralty would, on its behalf, undertake the task of arranging for the transfer and subsequent operation of the Establishment, and would, in due course, hand it over to the Ministry of Supply.

The activities of the Establishment fell into two main divisions:–

Pedigree seed development
I.   A problem of agricultural research which has been successfully developed by the Linen Industry Research Association, and which needed to be continued for development of new strains of pedigree seed and other cultural research. This work was centred at Fitcham Abbey under the general direction of the Research Association's Chief Botanist Mr G.O. Searle.

Fibre Production
II. A problem of industrial research hitherto limited to a small scale production of tank-retted fibre for wet spinning by the North Ireland Linen Industry. This had also been managed by the Association's Chief Botanist on directions formulated by a Committee representing both the Research Association, and the Company formed to operate the Flax Processing Factory located at West Newton. This factory is now being expanded to handle the crop from a much greater acreage of flax.

The meeting concluded:

It was agreed:
i.   That the extensions to the Factory now in hand should be completed before transfer to the Government, and that the 1st October should be worked to as the date for transfer of the Norfolk holdings to the Government.
ii. That the Linen Industry Research Association and Norfolk Flax Limited, should be noticed formally of the Government's intention and invited to formulate their proposals for effecting the transfer by

the 1st October.

iii. That the Association should be asked to release for service with the Government, Mr Searle and other officers and staff of the Association now attached for duty in Norfolk.

iv. That a joint approach by the Ministry of Supply and Admiralty should be made to the Treasury for the purpose of discussing staff proposals for running the Establishment as a Government concern on the general lines indicated in the discussion.

Furthermore, Gordon Herriot Cunningham FRS took a leading part in establishing a linen flax industry in New Zealand during the War. He spent most of his time travelling in the flax growing districts selecting suitable areas, advising growers on cultural disease control and organising seed supplies.[25]

Continuing the theme of vital, yet more unusual commodities, The Geological Survey which was founded in 1835 and mainly concerned with mapping geological deposits of minerals, turned its attention to other issues including the availability of underground water in addition to the water condition. In a paper initially published in the BWA Journal, Sir Edward Bailey FRS, the Director of the Geological Survey, titled *The Geological Survey in Relation to underground Water.* The abstract stated:

> The main object of the Geological Survey in relation to underground water has been, and still is, to collect, correlate and make available on a national scale existing, scattered, un-coordinated information.[26]

It took time, but was fruitful in identifying underground water sources.

Other Engineering Improvements

The following is a brief resume of some of the other technologies:

*   Adhesives specifically the development of REDUX (standing for Research at Duxford), used to bond thousands of Cromwell and Churchill tank clutch plates – Norman Adrian de Bruyne FRS.[27]
*   Aero engine design. One of the leading experts was – Lionel Haworth FRS.[28]
*   Aircraft design of the Brabazon specifically both aerodynamically and structurally – William John Strang FRS.[29]
*   Aircraft stressman at factories in Coventry and Reading – John Maynard Smith FRS.[30]
*   Construction of a plant for the extraction of magnesium hydroxide from the sea at Barry Docks, South Wales, a large underground factory for the manufacture of aeroplane engines. Supervision of a number of the

reinforced concrete caissons and Phoenix units that were part of Mulberry Harbour for the invasion of France. The larger units were built in excavated basins behind the left banks of the Thames below Barking, and when half completed the banks were cut, the units were floated out at high tide and towed to docks and the upper parts were completed – Sir (Thomas) Angus Lyall Paton FRS.[31]

- Creep-resisting alloys for gas turbine engines – Leonard Rotherham CBE FRS.[32]
- Investigation of propeller-induced vibration of the new Typhoon fighter resulting in anti-vibration mountings for the engine and a spring seat for the pilot – Sir Frederick William Page FRS.[33]
- Modified a Wellington bomber to operate as a destroyer of enemy magnetic mines – Sir George Robert Edwards OM FRS.[34]
- Powder metallurgy of the tungsten carbide/cobalt blend that was used in the production of high-velocity armour-piercing shot – John Anthony Hardinge Giffard FRS, 3rd Earl of Halsbury.[35]
- Provided facilities for industrial research and to encourage the application of successful results to the industry concerned – Henton Morrogh FRS.[36]
- Research work on lubricants and bearings in Melbourne – David Tabor FRS.[37]
- Scientific help to protect ships against magnetic mines – Louis Eugène Félix Néel ForMemRS.[38]
- Steel-backed copper-lead bearings for Rolls Royce Merlin engines used by the Royal Australian Air Force – Sir Robert William Kerr Honeycombe FRS.[39]
- Synthesis of hydrocarbons that improve the performance of gasoline – Kenichi Fukui ForMemRS.[40]
- Welding research at the Cambridge Engineering Laboratory – Michael Rex Horne FRS.[41]
- Wind tunnel experiments to calculate the effect of the walls of a wind tunnel of octagonal cross-section on the lift and drag forces on models mounted in the tunnel – George Keith Batchelor FRS.[42]

Additionally, Walter Thompson Welford FRS worked for the scientific instrument makers Adam Hilger Ltd in Camden Town (from 1942). Frank Twyman FRS, the distinguished optical engineer, was Managing Director.[43]

Sir William Arthur Stanier FRS is also noteworthy here. The following are comments from his Royal Society Biographical Memoir:

Stanier had not forgotten the War service of the Great Western and in 1934 he made a survey of all the machine tools on the L.M.S. for use in an emergency and he also reported its workshop capacity to the Committee for

Imperial Defence. In June 1937 the Master General of Ordnance, General Sir Hugh Elles, a friend from World War One, ask if the L.M.S. could undertake the design and construction of tanks (armoured fighting vehicles). Stanier agreed to this and a special design office was set up in London under H.G. Ivatt who was responsible for all subsequent tank programs carried out by the L.M.S.[44]

Figure 8.4 Sir William Arthur Stanier. (© Godfrey Argent Studio. Reproduced with permission)

9   Our health and wellbeing

There were far too many scientists and engineers involved in projects related to our general health and wellbeing to mention them all. The discussion will therefore focus on some of the issues relating to food, medical matters, chemicals, notably penicillin, and personal safety.

Food Supplies

The Agricultural Research Council met on many occasions during the War. Minutes from any of their meetings are interesting and demonstrate the range of issues facing the industry. Perhaps by way of an example some of the points raised during the sixty second meeting held on Tuesday 23 November 1943 could be used to demonstrate the scope of the Committee.

The following were present:

> The Right Honourable The Early De La Warr (in the Chair), Professor F.T. Brooks FRS, Professor I. De Burgh Daly FRS, Sir William Cecil Dampier FRS, Sir Frank Engledow FRS, Sir James Gray FRS, Sir Robert Greig, Sir Charles Harington FRS, Professor D. Keilin FRS, Sir Edward Salisbury FRS, John Smith, Dr W.W.C. Topley FRS (Secretary) and E.H.E. Havelock (Administrative Secretary). Mr V.E. Wilkins (Ministry of Agriculture and Fisheries) and Mr J.M. Caie (Department of Agriculture for Scotland) attended as Assessors to the Council.

> Sir Frank Engledow suggested that the Council should deal with the following matters, on which action was urgently required:-

> 1. A general survey of the present situation in agricultural research. Such a survey had been begun by groups appointed by the Joint Committee of the Agricultural Improvement Councils and the Agricultural Research Council.

2. The relations between Research Institutes and Universities.
3. Recruitment and training of research workers. Possibly it was desirable to establish one scheme of recruitment for teaching, research and perhaps advisory work.
4. The formulation of plans to ensure that proper arrangements are made for research work carried out by teaching staffs, both as regards finance and co-ordination with research work elsewhere.
5. The chain of connection between research and farmer, involving the repetition of initial experiments at first at four or five places and finally, in every County this was a matter which, to a large extent, could be dealt through the Joint Committee of the Agricultural Improvement Councils and the Agricultural Research Council.
6. Suitable salary scales for research workers.
7. Arrangements for consideration by the Council of the programs of research of the Institutes and their annual reports.
8. Procedure for keeping the Council in touch with general development in each branch of agricultural science.
9. Arrangements for the examination by the Council of applications for the special research grants.
10. Arrangements to relieve the Council of as much administrative and financial detail as possible.

Further in the meeting there was a discussion of two memoranda, in which all members of the Council present as well as the assessors took part. During this discussion the following were some of the chief matters to which attention was directed:

a) The importance of linking teaching and research.
b) The urgency of a decision on the major question whether the aim of future policy should be entrusted to the administration of Research Institutes to Universities or to independent governing bodies, dealing with the Ministry of Agriculture and Fisheries or Department of Agriculture for Scotland, and the Agricultural Research Council. The difficulties, both as regards detailed administration and general questions of policy, for example the closing of an Institute or its removal elsewhere, which arose when an Institute as a national centre of research, was owned and administered by a University, were noted. There was general agreement as to the desirability of closer scientific contacts between Institutes and Universities, which can be assisted by the interchange of staff and by allowing research workers to do some teaching, and there was full realisation of the fact that any agreed policy could not be applied at once or universally to existing Institutes.

c)   The need for great care in the choice of suitable men as Directors, and of preserving a large measure of freedom to Directors in regard to choice of program and general scientific responsibility.

d)   The desirability of devising as simple and speedy administrative mechanism as possible for dealing with all matters arising between the Government Departments concerned and the Institutes.

e)   The necessity for freeing Directors as much as possible from administrative detail by the appointment of competent secretaries to Institutes, so that Directors may take an active part in research.

f)   The desirability of ensuring that intimate contacts are developed between members of the Council and the staffs of the Institutes. While Committees of inquiry are on occasion necessary, the normal method should be that of informal visits by individual Council members.

g)   The vital importance of attracting able men into agricultural research. As regards junior workers, there should be propaganda in the schools, and Heads of Department at Universities should be informed that if they know of able young men, the Agricultural Research Council will be prepared to provide grants, according to the student's income, for their further training, which should usually include some technical training in agriculture and in other agricultural sciences than the one chosen for specialisation. The possibility of securing immediately after the War the services of young biologists now in the Forces or engaged on technical war-work should be explored thoroughly. More senior workers, including those whose experience has been in pure science, may be attracted to the Council's Research Unit or be appointed as individual members of the Council's scientific staff.

h)   The need for further consideration of the question whether certain of the smaller Institutes already administered by Universities should not in future be absorbed into Departments of Agriculture or Veterinary Medicine, adequate provision being set aside for research in the educational grants to those Departments.[1]

In common with other scientific organisations, the Agricultural Research Council had Sub-Committees one of which was the Horticultural Sub-Committee of the Agricultural Research Council and the Agricultural Improvement Council. The Sub-Committee minutes produced below predate the minutes of the Agricultural Research Council mentioned above, as they were chosen at random to reflect the issues of the day. The potential use of colonial surpluses lasted for a number of years.

Horticultural Sub-Committee of the Agricultural Research Council and the Agricultural Improvement Council

The constitution of this Sub-Committee was: Sir Edward Salisbury FRS (Chairman), Sir Ronald Hatton FRS, Dr T. Wallace FRS, Dr H.V. Taylor, Mr J.A. Symon, Mr D. Akenhead (Secretary).

The terms of reference of this Sub-Committee were:

to carry out a preliminary survey of the field of horticultural research in respect of fruit and vegetables, with the object of determining how best the research programs at Institutes, Advisory Centres, Universities or elsewhere can be extended or modified so as to ensure (a) a general advance in the application of basic scientific knowledge to the problems of horticulture, and (b) that the major practical problems (research on potatoes is receiving attention elsewhere) of the horticultural industry are attacked.[2]

This Sub-Committee undertook a survey of horticulture some of the data from which appears as figure 9.1.

Vegetables	Total acreage in 1941/42
Asparagus	1901
Beans (broad)	6681
Beans (French/runner)	8392
Beetroot	13,561[1]
Broccoli (heading)	26530
Brussels sprouts	38295
Cabbage (spring)	47155[2]
Cabbage (Autumn and winter)	70468[2]
Carrots	37251[3]
Cauliflower	17632
Celery	4373
Leeks	2352
Lettuce	8768[4]
Onions (salad)	1908
Onions (bulb)	9545
Parsnips	7104
Peas (canning)	18342
Peas (green)	50333

Vegetables	Total acreage in 1941/42
Peas (dry)	78478
Rhubarb	5937
Swedes/turnips (human consumption)	37855
Tomatoes (outside)	3292
Tomatoes (inside)	3045
Total	499198

Notes:
1. Total crop probably higher as some is harvested prior to September.
2. There may be overlapping in this total.
3. More of this crop is sown after June 4th.
4. This crop can only be estimated owing to overlap between returns.

Figure 9.1 Total acreage in England and Wales devoted to individual vegetable crops.

Whilst the United Kingdom ate all it produced, perhaps somewhat surprisingly there were some foods in overabundance in some parts of the world, not all of which might have been deemed palatable in the West. For example:

Dear Everett,

With your letter of the 5th November you sent me a copy of a letter dated 18th August from Suraj Kumar who writes from the Punjab on the subject of snake-meat.

I have consulted Chandler and enclose a copy of his memorandum dated 13th November, with which I am in agreement.

The only point which strikes me, and which doubtless will have struck your Chairman also, is that caste prejudices may have to be taken into consideration. The diet of Indian hill-tribes is not usually a safe precedent either for Hindus or for Moslems and great care would be necessary to reconcile dietary benefits with caste considerations.

However, these are matters which will doubtless be taken into consideration by the Punjab Government if your Chairman considers it suitable to advise Suraj Kumar to refer his suggestion to that Government.

Yours sincerely,
H.A. Lindsay

There were, however, more conventional foods in surplus.

Imperial Institute
Alternative Uses for Surplus Products
Egg Products
Survey of Present Position

The chief authorities consulted in this enquiry have been Dr F. Kidd FRS and Dr E.C. Bate-Smith of the Low Temperature Research Station, Cambridge (Department of Scientific and Industrial Research). Contact has also been made with Mr W.L. Arkinson Dairy Officer and Mr E. Macfarlane, Commonwealth Department of Commerce, Office of the High Commissioner for Australia, London; De W.H. Cook, Director, Division of Applied Biology National Research Council, Canada, through the Canadian Government Trade Commissioner in London; Messrs Armour and Company, Ltd, London; and Mr M. Wick and Mr Townsend of Messrs S.F. Wick and Sons Ltd, London. A visit was also paid to the works of Messrs Lactagol Ltd at Mitcham, to witness an experimental spray drying of eggs by the Kestner process, conducted by officers of the Low Temperature Research Station.

The report covered: Methods of Preparation; The Cambridge Investigations; Conclusions

In a letter to Lord Hankey FRS, in his capacity of Chairman of the Scientific Advisory Committee, Sir F.W. Leith-Ross commented (on 22 May 1941):

You also asked for a list of raw materials which we wish you to consider. The full list of commodities now being treated as in surplus supply is given below. Many of them are already mentioned in the documents previously sent to you, and others, of course, may not call for any immediate action:

Bananas	Citrus fruit	Cocoa	Coffee
Copra	Cotton	Jute	Lead
Maize	Meat	Milk products	Oilseeds
Pepper	Rice	Sisal	Sugar
Tanning extracts	Wheat	Wool	

Amongst the possibilities mentioned in the memoranda by the Imperial Institute, the following have been drawn to our attention:
- Drying of bananas.
- By-products from citrus fruits.
- Power alcohol from molasses.
- Oilseeds as a source of fuel oil.

Other possibilities which have been drawn to our attention are:

a)   The drying of butter fat.

b)   The improvement of whole cream milk powder.

c)   Egg drying.

d)   Extraction of vitamins from citrus fruit.

e)   Extraction of power alcohol from wheat and coffee.

f)   The research work now proceeding at New Orleans on alternative uses for cotton.

The representative of the India Office on the Official Surpluses Committee stated that research work was being conducted under the auspices of the Government of India, especially in respect of the production of acetone, yeast, groundnut oil, and new uses for cotton, and it was suggested that your Committee should arrange to keep in touch and to exchange information with the Government of India's scientific advisors.

My Committee is hardly equipped to give you immediate guidance on the research work which may be initiated or expedited on these products, or on the order of precedence which may be affixed to the work already in progress here and elsewhere on any particular commodity. Our desire can be described in the following terms:

•   To discover an alternative use for each product so that the surplus, or a portion thereof, can be utilised for some purpose other than the usual one, preferably by conversion into some product immediately useful for War purposes, without thereby creating a new surplus of a commodity which will not be less difficult to handle than the original surplus.

•   To discover processes which will improve ability to preserve the perishable products in store over a long period, and preferably processes which do not require the installation of complicated machinery which may also be difficult to provide or procure.

•   To apply in the country of origin, processes which will sensibly reduce the bulk of the product so that, without reducing its essential value, less shipping is required to transport it.

Yours sincerely

Sir Frederick Leith-Ross

There are other documents in the same National Archives folder relating to surplus foods such as cocoa and coffee.

The issue then became one of how to preserve the surplus. The meeting minutes for 25 April 1941 of the Scientific Advisory Committee (see Chapter 3 for further details of this Committee), recorded the following:

Alternative uses for surplus products

The Committee had before them the following papers:-
1. Note by the joint Secretaries covering correspondence between the Chairman and Sir Frederick Leith-Ross including notes by the Imperial Institute on certain crop surpluses.
2. Note by the Joint Secretaries covering the First Report of the Scientific Committee for examining Alternative Uses of Colonial Raw Materials.

The Chairman suggested that the Committee should examine first the recommendation put forward in the second of these two papers that a Research Institute should be created to conduct research into the properties of Colonial products of which there was a prospective surplus and the possibility of using them for the production of raw materials for industry. There were three aspects of this proposal which he would like the Committee to consider:
i.   Should the proposed organisation confine itself to research into Colonial products or should it be prepared to undertake investigation on behalf of the Dominions and India?
ii.   Was there any possibility that the work of the Institute would duplicate, or overlap with, that of existing organisations such as the Imperial Institute?
iii. If such an organisation were to be set up should it be attached to the Colonial Office or on the other hand to some existing organisation in the scientific field such as the Department of Scientific and Industrial Research?[3]

Some months later a paper was published.

Food Investigation
University of Cambridge and
Department of Scientific and Industrial Research

Low Temperature Research Station
Downing Street, Cambridge

Interim record memorandum no 125

Subject: Dehydrated food and transport, with particular reference to Food Supplied for the fighting forces and for emergency feeding under War conditions.

Authors: F. Kidd FRS and T.N. Norris
Date: 7th January 1942

1.   In Interim Record Memorandum No 83 Dehydrated Foodstuffs and
     Transport it was pointed out that dehydrated foodstuffs do not require
     refrigeration during storage and transport. Secondly, their weight and
     volume are far less than those of the corresponding fresh and canned
     foodstuffs. It has also been established that dried foodstuffs can today
     be produced of excellent quality as regards both palatability and nutri-
     tive value. Impressions based on experience of dried products in the
     past are no longer valid.
2.   Table 1 shows the calories per pound of various kinds of food in the
     fresh, canned and dried food, which would be required for a full
     (messing) ration and also the calories that would be carried per cubic
     foot of storage space. Table 2 gives an example of a ration providing
     3,500 calories per day and consisting of flour, biscuits, other cereals,
     potatoes, carrots, pulses, meats, soups, fats, sugar, fish, eggs and milk.
     From this example it appears that the volume of foods required (as
     packed for storage and transport), in the dried form is, as compared
     with the canned form, approximately 1 to 4 while the comparison in
     weight is similar. For the same transport facilities therefore one could
     probably carry four times the amount of the various foodstuffs required
     for a full ration of dried food as compared with canned or fresh food.
3.   While research has shown that these possibilities are practicable,
     commercial production to enable them to be realised on a large scale
     would require careful organisation and must take some time. In the
     development from laboratory scale to full commercial production, a
     series of problems must be expected to be met, all of which will be
     soluble but will require a shorter or longer time for solution. Large scale
     manufacture can rarely be undertaken with success at short notice.[4]

Clearly these two reports are not connected, but are both related to food
and how to preserve or utilise available supplies.

Of course there were fulltime research organisations with the necessary
resources and provided scientists to undertake some of the necessary work
– not all of it was conducted at universities. For example the Rothamsted
Experimental Station was founded in 1843 on the appointment of a chemist
Joseph Henry Gilbert by the owner of the Rothamsted Estate, Sir John
Bennet Lawes. This scientific partnership lasted 57 years lying the foundations of modern scientific agriculture and established the principles of crop
nutrition. This organisation still exists today.[5]

Then as now there were some very talented scientists working at the Rothamsted Experimental Station. For example John Burdon Sanderson Haldane FRS moved there in 1941 and stayed until 1944.[6] Whilst the Rothamsted Experimental Station concentrated on agriculture, the East Malling Research Station was, and still is, concerned with the culture of fruit trees and bushes. Sir Ronald George Hatton FRS was Director of the East Malling Research Station for thirty years (1919–1949).[7]

Other Fellows held various posts, both academic and Government, relating to agriculture during the War, see below:

Name	Occupation
Ford (Charles Edmund)	Worked as geneticist for the Rubber Research Scheme in Celon (1938–1945) with an interruption for War service in 1942–43 as a Lieutenant in the Royal Artillery.[8]
Fryer (Sir John Claud Fortescue)	Secretary the Agricultural Research Council (from July 1944).[9]
Kidd (Franklin)	Worked on the dehydration of milk and dried foods.[10]
Marston (Hedley Ralph)	Officer in Charge of the Nutrition Laboratory of the CSIR (the Commonwealth Council for Scientific and Industrial Research Adelaide South Australia). The Laboratory was re-established as a Division of the CSIR (Biochemistry and General Nutrition) in August 1944 with Marston as Chief.[11]
Neuberger (Albert)	Worked on the nutritional value of the potato. He was joined in this work by a newly graduated student called Fred Sanger FRS (subsequently a double Chemistry Nobel laureate), who became his first PhD student.[12]
Orr (John Boyd, 1st Baron Boyd-Orr of Brechin Mearns)	Professor of Agriculture at the University of Aberdeen (1942–45). He also took part in discussions on a World Food Plan.[13]
Robertson (Sir Rutherford Ness)	Australian scientist who developed procedures to prolong the storage life of apples and other fruit and to the open-air storage of the huge reserves of wheat in Australia.[14]
Russell (Sir Edward John)	Retired from the post of Director of Rothamsted in 1943 (after 31 years working there).[15]
Salaman (Redcliffe Nathan)	Section Commander in the Home Guard. He also acted as Billeting Officer, Voluntary Food Organiser and Member of the Potato Advisory Committee, Ministry of Agriculture.[16]
Salt (George)	Studied wireworm ecology which were devastating when pastures were ploughed and sown to wheat and other crops.[17]
Wain (Ralph Louis)	Began work at Long Ashton Research Station with Dr E.H. Wilkinson (Sept. 1939) on the mode of action of copper fungicides and later potential insecticidal including an interest in DDT ('dichlorodiphenyltrichloroethane') and related substances after which he studied plant growth hormones.[18]

Name	Occupation
Wallace (Thomas)	Head of the Unit of Plant Nutrition (Micro-nutrients) within the ARC.[19]
Yule (George Udny)	Director of Requirements in the Ministry of Food.[20]

Figure 9.2 Some Fellows involved in Food Research or administration.

Animal nutrition was also important to the food chain. Figure 9.3 details some of those individuals involved in animal welfare.

Name	Occupation
Brooksby (John Burns)	Assisted Sir William Henderson FRS during Gregor's development of a method for titrating the foot-and-mouth virus.[21]
Coombs (Robert Royston Amos (Robin) Coombs)	Directed to the Ministry of Agriculture's Veterinary Research Laboratory at Weybridge (1943) to work with Norman Hole, the head of the diagnostic section, on the sero-diagnosis of glanders, a fatal disease of horses caused by Burkholderia mallei, which was considered a possible biological warfare agent and was so used by the Japanese.[22]
Hammond (Sir John)	Senior Assistant, Animal Nutrition Institute.[23]
Henderson (Sir William Macgregor)	Joined the staff of the Animal Virus Research Institute (AVRI), Pirbright (1938) where he worked on foot-and-mouth disease.[24]
Rowson (Lionel Edward Aston)	Appointed Sterility Investigation Officer for the Ministry of Agriculture (1940) and worked in the Pathology Laboratories on the Downing site.[25]

Figure 9.3 Some Fellows involved in animal welfare or administration.

Medical projects

Section 5.4 documented some of the Fellows who occupied medical posts during the War. This section will concentrate on some of the projects/ programs whose output was directly intended to increase the wellbeing of either civilian or Services personnel.

- **Artificial lenses for human eyes** – Sir Nicholas Harold Lloyd Ridley FRS. On his flight back from the 'Adlertag' ('Day of the Eagle') battle on 14 August 1940, a bullet smashed through the Perspex acrylic material that formed the sidewalls of Flight Lieutenant Gordon 'Mouse' Cleaver's cockpit.

Unfortunately he was flying without goggles having forgotten them prior to take off. He was immediately blinded in both eyes by multiple fragments of plastic from the canopy. There followed a series of operations performed by Sir Nicholas on Cleaver's eye over the next several years at Moorfields. Sir Nicholas' Royal Society Biographical Memoir offers the following of these events: 'some of these operations consisted of removing pieces or portions of the plastic fragments embedded in the coats surrounding his eyes or within the eyes. One eye was saved and he was eventually able to return to civilian life with several small pieces of plastic still embedded in his eyes. What, then, did Flight Lieutenant Cleaver's injuries have to do with the invention of the IOL more than eight years later? Harold had long believed that it would be possible to replace cataractous lenses with artificial lenses … That realization led Harold to believe that biocompatible artificial lenses could be made of a similar material to replace the natural lenses, thus restoring much improved vision to patients with cataracts.'[26]

- **Bacteriology** – Martin Rivers Pollock FRS 'was involved in providing bacteriology in an isolation hospital Leicester – typically patients with diphtheria, haemolytic streptococci, salmonella or typhoid, and being prepared for any crises that would result from bacteriological warfare.'[27]
- **Blood donors** – one of the items in the report on the Sub-Committee appointed by the Transfusion Research Committee of the Medical Research Council considered the question of supplying a badge or certificate for blood donors. The report contains the following comment: 'many donors have asked for a badge. They can be worn in the street and in the case of older people especially would give them the satisfaction of knowing their form of War Service was recognised.'[28]
- **Blood serum** – Ralph Ambrose Kekwick FRS 'worked at the Lister Institute investigating the practical problems connected with the preparation of serum and plasma for transfusion.'[29]
- **Brain damage** – Oliver Louis Zangwill FRS 'assessed brain damage in the Brain Injuries Unit in Edinburgh.'[30]
- **Contact dermatitis** – Philip George Houthem Gell FRS 'researched contact dermatitis which was a serious problem among workers in ordnance factories who often developed allergic skin reactions to the chemicals in explosives.'[31]
- **Dried blood plasma** – Ruth Ann Sanger FRS 'took a post on the scientific staff of the NSW Red Cross Blood Transfusion Service (Director, R.J. Walsh) initially to work on drying blood plasma.'[32]
- **Human infections** – Sir Alexander Fleming FRS (awarded the Physiology or Medicine Nobel Prize in 1945) 'worked on the treatment of a few severe human infections.'[33]
- **Louse control** – Patrick Alfred Buxton FRS 'devoted the main effort of his

department to the improvement of practical measures against the louse.'[34]

- **Mosquito control** – Douglas Frew Waterhouse FRS 'worked in Canberra with the commissioned rank of Captain in the Army Medical Corps (on the Officer Reserves). He tested high-spreading oils for mosquito control, but it soon evolved into the testing sprays for control of mosquitoes responsible for malarial transmission.'[35]

- **Neurology service** – Sir Gordon Morgan Holmes FRS 'set up a neurology service whilst a consultant to the Emergency Medical Service.'[36]

- **Pulmonary diseases** – Claude Gordon Douglas FRS 'initiated and supported a large-scale investigation into the incidence, nature and causes of the disabling pulmonary diseases of Welsh colliers. He was also Chairman of the Medical Research Council's Committee on Industrial Pulmonary Diseases (from 1936), Member of the Chemical Board of the Ministry of Supply (1939–46) and Chairman of some of its Sub-Committees.'[37]

- **Rationing** – as soon as World War Two began, Elsie May Widdowson CH FRS working with someone called McCance started an experimental study of rationing using themselves and their friends as test subjects. Their basis theme was to reduce as much as possible the amount of food needed to live. Their critics felt that such low quantities of eggs, meat, milk and other foodstuffs were so small that they were considered intolerable by their critics.[38]

- **Rickets** – Sir Edward Mellanby FRS 'experimental study of the metabolic defect in rickets. Sir Edward was also active in many Government Committees.'[39]

- **Tuberculosis** – Gerhard Domagk ForMemRS (awarded the Physiology or Medicine Nobel Prize in 1939) 'was chiefly engaged in a chemotherapeutic approach to tuberculosis.'[40]

- **Water/human physiology** – Brian Blundell Boycott FRS 'was concerned with a search for drugs to ameliorate seasickness among troops making amphibious landings. He also worked on problems of diving physiology, particularly the effects of oxygen at high pressure and carbon dioxide narcosis.'[41]

Chemicals and pharmaceuticals

Whilst some of the chemistry research was undertaken in university chemistry laboratories, there was a central Chemical Research Laboratory, mentioned in figure 1.5 as reporting to the Department of Scientific and Industrial Research (DSIR). The potentially large sites of chemical research necessitated a co-ordinated approach so that everyone, from industry, university or the DSIR was aware of each other's potential research interests – at least in broad terms.

Accordingly, a confidential paper (CRD paper no 211) was issued by the Chemistry Research Board (a body reporting to the DSIR), titled *The Functions and Future Program of the Chemical Research Laboratory*.

A summary of the document details the following points:

1) The functions of the Chemical Research Laboratory are first examined and the general considerations underlying its choice of program are given.

2) Main themes of the Laboratory's program should be: the collection of reference data; the study of new techniques; the conservation of essential materials; the utilisation of indigenous raw materials (including wastes) and the examination of new chemical processes and materials. These themes are exemplified.

3) Recommendations in general terms are made on the general future trend of the Laboratory's research work. The future of its Microbiological research and of development work and chemical engineering research requires further examination.

4) The role of the Laboratory in carrying out work on request for other agencies, and in advisory work are discussed.

5) General observations on the size of the Laboratory are made.

6) It is recommended that the Laboratory be considerably increased in size as a matter of urgency. An approximate doubling should be the first step. A preliminary estimate of the requirements in staff, buildings and money are given.

7) It is recommended that the functions of the Laboratory (last defined in 1938) be redefined as follows:

   a) To carry out fundamental chemical research on behalf of the Department of Scientific and Industrial Research including development work so far as this may be necessary to demonstrate the industrial value of any discoveries which may be made.

   b) To carry out appropriate chemical research on request and to provide technical advice to other Sections of the Department, other Government Departments and British Industry.[42]

The report goes on to mention examples of new materials and processes which are used here as examples of the types of chemical technology under development in this group through the War.

   a) **Plastics and high polymers** – this is a field in which the Laboratory has been engaged since its foundation, and it had a useful record of achievement including at least one major discovery to its credit. High polymers and plastics today constitute an important section the materials of technology.

b) **Microbiology and applied chemistry** – on the recommendation of a
Panel consisting of Professors A.C. Chibnall FRS and H. Raistrick
FRS and Dr F.H. Carr, the following topics were selected.

Chemical work other than through this organisation and already
mentioned, involved work on the atomic bomb and some new explosives.
One compound not previously mentioned, which was the subject of much
work both in the United Kingdom and the United States was penicillin. Sir
Alexander Fleming FRS, Sir Ernst B. Chain FRS and Lord Howard Florey
FRS were jointly awarded the Nobel Prize in Physiology or Medicine for 1945
with the citation:

> for the discovery of penicillin and its curative effect in various
> infectious diseases[43]

All three chemists were very active during the War. Arguably Lord Florey
FRS, (see figure 9.4), had a higher international role than the other two
scientists.

Howard Walter Florey was born in Adelaide, Australia on the 24 September
1898 and died in Oxford on 21 February 1968. Some of the highlights of his
career were:

- Adelaide University medical school (MB, BS), Oxford University (BSc),
  Cambridge University (PhD 1927) Department of Pathology, Cambridge
  University (1924).
- Rockefeller Scholar, studying microsurgical techniques (1925).
- Freedom Research Fellow, London Hospital (1926).
- Huddersfield Lecturer in Pathology, University of Cambridge (1927).
- Fellow, Gonville and Caius College.
- Director of Medical Studies, Gonville and Caius.
- Joseph Hunter Professor of Pathology, University of Sheffield (1932–1935).
- Professor of Pathology, University of Oxford (1935–1962).
- Fellow, Lincoln College, Oxford (1935).

Penicillin has a remarkably simple chemical structure, (see figure 9.5), at
least superficially.

Figure 9.5 is not strictly one chemical. The 'R' symbol is used by chemists
to denote that different chemical groups might be attached to this position. It
is used above to show that following the first successful synthesis of the drug,
other variations were the subject of further research, however that is a much
later story.

Understanding the chemical structure and a means by which the material
could be scaled up to sufficient quantities was a long story involving many

Figure 9.4 Lord Florey (© Godfrey Argent Studio. Reproduced with permission)

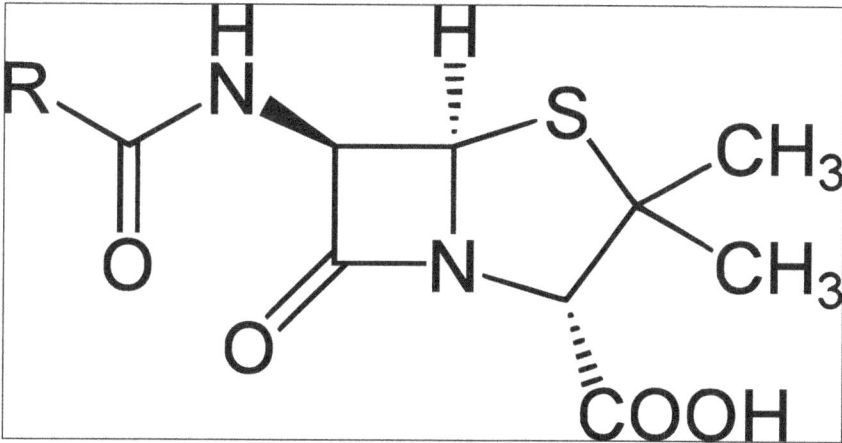

Figure 9.5 The structure of penicillin. (Author's diagram)

scientists, some of which had already been awarded the Nobel Prize, or went on to the award later. They included:

- Lord Alexander Todd OM FRS (Chemistry Nobel Prize 1957) 'became a member of the ICI Dyestuffs Research Panel, an important member of the anti-malarial project, a member of the international penicillin project and worked on chemical warfare agents. Incidentally, his father in law was Sir Henry Dale OM PRS (the Wartime President of the Royal Society and himself a Nobel Laureate).'[44]

- Sir Robert Robinson OM FRS (Chemistry Nobel Prize 1957) 'was involved in a variety of Government activities and served on many bodies concerned with the chemical defence and explosives and, through the Medical Research Council, chemotherapy. He was the Waynflete Professor of Chemistry at Oxford University and also researched chemical warfare agents and anti-malarial drugs for example. Robinson went on to be President of the Royal Society from 1945–1950.'[45]

- Robert Burns Woodward ForMemRS (Chemistry Nobel Prize 1965). 'During the period of World War Two, Woodward played an important part in the joint Anglo-American effort on penicillin and he was indeed the first to postulate the correct structure of that antibiotic.'[46]

- Howard Walter Florey (Baron Florey of Adelaide and Marston) OM FRS (Physiology or Medicine Nobel Prize 1945) 'studied penicillin throughout the War. His visit to the United States in 1943 led to a remarkable effort by the American firms resulting in the production of enough penicillin to treat all severe casualties by D-Day in Normandy in 1944.'[47]

- Dorothy Mary Crowfoot Hodgkin OM FRS (Chemistry Nobel Prize 1964) 'was an x-ray crystallographer credited with determining the structure of penicillin. She later worked with Professor Bernal FRS mentioned in earlier chapters. Incidentally, Hodgkin was the first and is still the only British female Chemistry Nobel Laureate!'[48]

Dorothy Mary Crowfoot Hodgkin OM FRS, (see figure 9.6), led an interesting life. Written by Georgina Ferry, Hodgkin's biography, titled *Dorothy Hodgkin A Life*, documents her early life in Cairo through to her role as a world expert in x-ray crystallography.[49] This technique was used by Hodgkin to understand the structure of penicillin and later the structures of some other natural products such as insulin and vitamin B_{12}.

Ferry included the following quote made by Hodgkin on one of her early encounters with Sir Robert Robinson (who also appears on the above list of Nobel Laureates and also played an important wartime role in penicillin research and development):

Figure 9.6 Dorothy Hodgkin. (© Godfrey Argent Studio. Reproduced with permission)

I was scared stiff talking to Robinson, and he started awfully badly by saying 'of course ICI have been financing a great deal of x-ray work lately ... I have always thought it a great waste of money myself'. My heart sank like anything, and then he suddenly said 'But your work and Bernal's is quite different. I have wanted it to be done in my department for a long time. What kind of room do you need? How much money?' ... my breath was quite taken away.[50]

Little did either realise they would both work on the same penicillin project during the War. Sadly, Hodgkin was confined to a wheelchair in later life and Sir Robert went blind.

By way of examples three of the other chemists involved in penicillin war-work were:

- Harold Raistrick FRS 'was appointed Demonstrator at the London School of Hygiene and Tropical Medicine in 1936 (and Lecturer in 1947). Served on the Medical Research Council Bacteriological Committee in 1938 and the Chemical Research Board and Flax Committee of the Department of Scientific and Industrial Research. He was the Scientific Adviser to the

Ministry of Supply on Penicillin Production 1944–47 and a member of their General Penicillin Committee.'[51]

- Ralph Raphael FRS 'was seconded to May and Baker to work on the chemistry and synthesis of penicillin.'[52]
- Wilson Baker FRS 'was one of the team of four at Oxford investigating the chemistry of penicillin.'[53]

On a related chemical theme, Donald Devereux Woods FRS was offered a Fellowship to work with Sir Paul Fildes FRS (a medical bacteriologist) in London (in 1939). On his arrival on 1 April 1939 the subject of sulphanilamide was under discussion with the available fact suggesting it inhibited bacterial growth.[54]

Some examples of other chemicals which received attention during the War are:

- **Blood group antigens** – Walter Thomas James Morgan FRS 'investigated the chemical structure of the blood group antigens and the substances responsible for their antigenic specificity.'[55]
- **Possible herbicides** – Sir James Baddiley FRS 'was one of 16 researchers who accompanied Lord Todd FRS when he moved to Cambridge (in 1944). An ICI Fellow he studied the synthesis of adenosine-5'-phosphate.'[56]
- **Rocket propellants and artificial antibodies** – Linus Pauling ForMemRS (Chemistry Nobel Prize (1954), Peace Nobel Prize (1962)) 'worked mainly on rocket propellants and in the search for artificial antibodies. Earlier he had used the paramagnetism of oxygen to design and develop an oxygen meter for use in submarines.'[57]
- **Rubber starting materials** – Arthur Charles Neish FRS 'joined the research laboratories of the National Research Council as a Research Officer, Division of Applied Biology (in 1943), one of a team with Dr Gordon Adams and Dr G.A. Ledingham working on the production by fermentation of 2,3-butanediol.'[58]
- **Sugar alternatives** – Stanley Peat FRS 'was initially involved in the study of uranium compounds Peat switched to sugar chemistry, particularly the production of glucose directly from potatoes. He became a member of the Cellulose and Cordite Panel of the Ministry of Supply, and his work for the Committee charged with finding a substitute, based on British seaweeds, for Japanese agar.'[59]
- **Toluene for explosives manufacture** – Dalziel Llewellyn Hammick FRS 'investigated the synthesis from benzene using a series of gas-phase reactions.'[60]

Personal Safety

In times of War personal safety takes on an urgency not thought of in peace time. Many Fellows were involved either in their own safety or belonged to organisations such as the Local Defence Volunteers for example. Others were involved as Fire Wardens. There were some Fellows, however, who worked on the personal safety of many individuals. Two examples of which are:

- George Ingle Finch FRS 'acted a Scientific Advisor to the Fire Research Division of the Ministry of Home Security, of which Lord Falmouth was head.'[61]
- William Joseph Elford FRS. 'From the beginning of the War in 1939 Elford was involved in work on airborne infection, both from the point of view of protection against bacteriological warfare and because of the danger of epidemics of air-borne disease in crowded air-raid shelters.'[62]

10   Conclusion

The post War years saw many changes to both Britain as a whole and the individuals mentioned above. Some of the scientists and engineers resumed their careers, some rose even further in their international standing, others retired. Throughout all of the national and individual changes, the Royal Society continued on its path, as it has for over 350 years, promoting excellence in science and engineering. With the exception of Sir Alan Lloyd Hodgkin OM FRS, one of the world's foremost neurophysiologists, and incidentally a Nobel Prize Laureate of Physiology or Medicine (1963), all of the Presidents of the Royal Society from 1950 to 1990 are mentioned in the previous chapters. They, along with the Presidents mentioned in figure 1.3 in Chapter 1, all undertook War related work, some with more prominence than others. They certainly could not be accused of remaining in ivory towers during the War. For completeness, figure 10.1 details these individuals.

Name	President for the years	Discipline
Adrian (Edgar Douglas, 1st Baron Adrian of Cambridge)	1950–1955	Physiologist
Hinshelwood (Sir Cyril Norman)	1955–1960	Chemist
Florey (Howard Walter, Baron Florey of Adelaide and Marston)	1960–1965	Pathologist
Blackett (Patrick Maynard Stuart, Baron Blackett of Chelsea)	1965–1970	Nuclear Physicist
Todd (Alexander Robertus, Baron Todd of Trumpington)	1975–1980	Chemist
Huxley (Sir Andrew Fielding)	1980–1985	Cell biologist
Porter (George, Baron Porter of Luddenham)	1985–1990	Chemist

Figure 10.1 Some post War Presidents of the Royal Society.

One might be tempted to assume that public recognition for war-work would be made after the event. Figure 10.2 details just the Fellows mentioned above along with national awards recognising their contribution to the War. Of course this list is incomplete. This account of the Fellows only includes some of them, thereby excluding many Fellows who gave their time and talents. Furthermore there are living Fellows, some of which prefer to live quiet lives away from the public gaze, who were also recognised for their efforts. Their story is yet to be told. One might assume that honours awarded for 1939, if not 1940, were for work undertaken in the pre-war years. They are included here for completeness. Furthermore, honours presented after 1950 may well have been for work undertaken after the War.

It is a further assumption, which has some credibility, that honours awarded for the intervening period related to war-work.

Name	Honour
Adrian (Edgar Douglas, 1st Baron Adrian of Cambridge)	OM 1942
Akers (Sir Wallace Alan)	CBE 1944, Kt 1946
Appleton (Sir Edward Victor)	KCB 1941, GBE 1946
Bailey (Sir Edward Battersby)	Kt 1945
Baker (John Fleetwood, Baron Baker of Windrush, Gloucestershire)	OBE 1941
Bartlett (Sir Frederic Charles)	CBE 1941, Kt 1948
Bhatnagar (Sir Shanti Swarupa)	Kt 1941
Boyd-Orr (John, 1st Baron Boyd-Orr of Brechin Mearns)	Baron 1949
Bragg (Sir William Lawrence)	Kt 1941
Bridges (Edward Ettingdean, Baron Bridges)	KCB 1939, GCB 1944, GCVO 1946
Brooks(Frederick Tom)	CBE 1947
Brown (Sir George Lindor)	CBE 1947
Bruce (Stanley Melbourne, Viscount Bruce of Melbourne)	Viscount 1947
Brunt (Sir David)	Kt 1949
Butler (Sir Edwin John)	Kt 1939
Buxton (Patrick Alfred)	CMG 1946
Chadwick (Sir James)	Kt 1945
Christopherson (Sir Derman Guy)	OBE 1946
Churchill (Sir Winston Leonard Spencer)	OM 1946
Coales (John Flavell)	OBE 1945
Cockcroft (Sir John Douglas)	CBE 1944, Kt 1948
Collingwood (Sir Edward Foyle)	CBE 1946
Cunningham (Gordon Herriot)	CBE 1949
Dale (Sir Henry Hallett)	GBE 1943, OM 1944

Name	Honour
Darwin (Sir Charles Galton)	KBE 1942
Dorey (Stanley Fabes)	CBE 1946
Dudley (Sir Sheldon Francis)	CB 1940, KCB 1942
Edgell (Sir John Augustine)	KBE 1942
Edwards (Sir George Robert)	MBE 1945
Egerton (Sir Alfred Charles Glyn)	Kt 1943
Engledow (Sir Frank Leonard)	Kt 1944
Ellis (Sir Charles Drummond)	Kt 1946
Everett (Douglas Hugh)	MBE 1946
Fairley (Neil Hamilton)	CBE 1941
Farren (Sir William Scott)	CB 1943
Fildes (Sir Paul Gordon)	Kt 1946
Fleming (Sir Alexander)	Kt 1944
Florey (Howard Walter, Baron Florey of Adelaide and Marston)	Kt 1944
Forster-Cooper (Sir Clive)	Kt 1946
Fowler (Sir Ralph Howard)	Kt 1942
Fox (Sir John Jacob)	Kt 1944
Frank (Sir Frederick Charles)	OBE 1946
Fryer (Sir John Claud Fortescue)	KBE 1946
Gates (Sidney Barrington)	OBE 1943
Glanville (Sir William Henry)	CBE 1944
Goldsbrough (George Ridsdale)	CBE 1948
Gough (Herbert John)	CB 1942
Gray (Sir James)	CBE 1946
Griffith (Alan Arnold)	CBE 1948
Guy (Sir Henry Lewis)	CBE 1945, Kt 1949
Hammond (Sir John)	CBE 1949
Hankey (Maurice Pascal Alers, 1st Baron Hankey of the Chart)	Baron 1939
Harington (Sir Charles Robert)	Kt 1948
Hartley (Sir Harold Brewer)	KCVO 1944
Hatton (Sir Ronald George)	Kt 1949
Heilbron (Sir Ian Morris)	Kt 1946
Hey (James Stanley)	MBE 1945
Hill (Archibald Vivian)	CBE 1946, CH 1946
Hinshelwood (Sir Cyril Norman)	Kt 1948
Hirst (John Malcolm)	DSC 1945
Jones (Sir Bennett Melvill)	Kt 1942

Name	Honour
Jones (Reginald Victor)	CBE 1942, CB 1946
Kay (Herbert Davenport)	CBE 1946
King (William Bernard Robinson)	MC 1940
Krishnan (Sir Kariamanikkam Srinivasa)	Kt 1946
Leiper (Robert Thomson)	CMG 1941
Lennard-Jones (Sir John Edward)	KBE 1946
Lindemann (Frederick Alexander, Viscount Cherwell)	Baron 1942
Lockspeiser (Sir Ben)	Kt 1946
Marshall (Sir Guy Anstruther Knox)	KCMG 1942
Mellanby (Sir Edward)	GBE 1948
Merton (Sir Thomas Ralph)	Kt 1944
Mountbatten (Louis Francis Albert Victor Nicholas, 1st Earl Mountbatten of Burma)	DSO 1941, CB 1943, KCB 1945, KG 1946, PC 1947, GCSI 1947, GCIE 1947
Paterson (Sir Clifford Copland)	Kt 1946
Pfeil (Leonard Bessemer)	OBE 1947
Pugsley (Sir Alfred Grenville)	OBE 1944
Ricardo (Sir Harry Ralph)	Kt 1948
Robinson (Sir Robert)	Kt 1939, OM 1949
Rothschild (Nathaniel Mayer Victor, 3rd Baron Rothschild)	GM 1944
Salisbury (Sir Edward James)	CBE 1939, Kt 1946
Schonland (Sir Basil Ferdinand Jamieson)	CBE 1948
Shoenberg (David)	MBE 1944
Simon (Sir Franz Eugen)	CBE 1946
Simonsen (Sir John)	Kt 1949
Smith (Sir Frank Edward)	GBE 1939, GCB 1942
Smuts (Jan Christian)	OM 1947
Southwell (Sir Richard Vynne)	Kt 1948
Spencer Jones (Sir Harold)	Kt 1943
Stapledon (Sir Reginald George)	Kt 1939
Steacie (Edgar William Richard)	OBE 1946
Stradling (Sir Reginald Edward)	Kt 1945
Stopford (John Sebastian Bach, Baron Stopford of Fallowfield)	Kt 1941
Storey (Harold Haydon)	CMG 1948
Taylor (Sir Geoffrey Ingram)	Kt 1944
Tizard (Sir Henry Thomas)	GCB 1949
Thomson (Sir George Paget)	Kt 1943

Name	Honour
Turing (Alan Mathison)	OBE 1946
Uvarov (Sir Boris Petrovich)	CMG 1943
Watson-Watt (Sir Robert Alexander)	CB 1941, Kt 1942
Whittle (Sir Frank)	CBE 1944, CB 1947, KBE 1948
Whittaker (Sir Edmund Taylor)	Kt 1945

Figure 10.2 Some Fellows honoured during the period 1939–1949.

Moving Forward

Mention was made at the start of Chapter 2 of some of the war-work of the Royal Society. Some of the Committees mentioned in that list appear in bold type, and have been discussed in detail in earlier Chapters except for the following four Committees:

- Advisory Committee on Airborne Research Facilities.
- Advisory Committee on Naval Research Facilities.
- Cultural Relations with India.
- Empire scientific conference.

These four Committees were active during the War, however, the output from their deliberations related to post War activities.

Advisory Committee on Airborne Research Facilities

Notes and Records provides the following overview for this Committee:

> Acting on a proposal that aircraft could be very profitably used for promoting scientific research in various fields, such as the control of food pests and insect vectors of disease, physical and biological metrology, oceanography, geophysics, movements and migrations of mammals, birds, fish and insects, air surveys of land forms, and archaeological surveys, the Council of the Society approached the Admiralty and the Air Ministry.[1]

A meeting of this Committee took place on 30 January 1946, with some familiar names:

> Sir Henry Tizard FRS, DR H.J.J. Braddick (representing Professor P.M.S. Blackett), Sir Edward Bullard FRS, Dr J.A. Carroll (Admiralty), Professor S. Chapman FRS, Mr R.C. Chilver (Air Ministry), Captain K.S. Colquhoun

(Admiralty), Dr G.B. Dickens (Air Ministry), Dr G.M.B. Dobson FRS, Sir Alfred Egerton FRS (Secretary R.S.), Mr G.H. Hollingdale (Ministry of Aircraft Production), Brigadier M. Hotine (War Office), Colonel G. Leitch (War Office), Dr A.C. Menzies (Deputy Chairman), Sir Thomas Merton FRS (Treasurer R.S.), Professor W.H. Pearsall FRS, Sir Frederick Russell FRS, Sir Edward Salisbury FRS (Secretary R.S.), Air Vice-Marshall T.M. Williams (Air Ministry), Rear Admiral A.G.N. Wyatt (Admiralty).

This Committee had several Sub-Committees including:

- The Aerial Magnetic Survey Sub-Committee.
- The Biological Sub-Committee.

Advisory Committee on Naval Research Facilities

The constitution of this Committee was:

Sir Edward Bullard FRS (in the chair), Vice-Admiral Sir John Edgell, Sir Alfred Egerton FRS, Captain L.G. Garbett, Sir Harold Spencer Jones FRS, Professor J. Proudman FRS, Dr C.S. Wright and Real-Admiral A.G.N. Wyatt.

With the following terms of reference:

To consider proposals for the use of naval facilities and personnel for assisting scientific research, and to make recommendations to the Council of the Royal Society and the Lords Commissioners of the Admiralty.[2]

The National Archives have some records of this Committee which shows a little of its work.

The Royal Society
Burlington House
London W1
30 October, 1945

Dear Admiral Wyatt,

Advisory Committee on Naval Research Facilities

You will have heard that the Council of the Royal Society accepted proposals which we made as a result of the informal meeting on naval research facilities held on 26th September 1945 at the Royal Society.

A request for such facilities has now reached me from Dr Bullard, Director of the Geophysical Laboratory Cambridge. He writes: ...

We have now heard from Meinesz that he is willing to lend his submarine gravity apparatus to us next year ...

He asks me to invite the Admiralty to arrange for a submarine to be available next summer for carrying out the program approved in 1938, west of the British Isles, and to find out the date when the work could commence. June would be suitable to Dr Bullard but a month or so either way does not matter to him.

I write to you to ask if you would be so good as to set going the preliminary arrangements necessary in order that the Admiralty may agree to allocate a submarine for this work, and I shall be glad to supply any further particulars which may be required.[3]

Cultural Relations with India

Little information could be found in the National Archives regarding this Committee. However, there is a list of members:

Sir David Ross	Chairman
Mr. T.P. Tunnard-Moore	Secretary
Dr D.A. Alann	(Museums Association)
Mr. K. De B. Codrington	(Victoria and Albert Museum)
Sir Lewis Fermor FRS	(Royal Society)
Professor H.J Fleure FRS	(Royal Anthropological Institute)
Professor H.A.R. Gibb	(Oxford University)
Sir Angus Gillan	(British Council)
Mr. B. Gray	(British Museum)
Lady Hartog	(East India Association)
Professor A.V. Hill FRS	(Royal Society)
Mr. A. Master	(School of Oriental and African Studies)
Mr. F. Richter	(Royal India Society)
Sir Theodore Tasker	(Royal Anthropological Institute)
Mr. T.G. P. Spear	(Cambridge University)
Sir John Simonsen FRS	(British Association)
Professor F.W. Thomas	(British Academy)
Professor R.L. Turner	(London University)
Mr. R.B. Littlehailes	(Universities Bureau of the British Empire)
Sir George Schuster	
Dr S.G. Vesey Fitzgerald	(London University)
Mr. E. White	(Arts Council of Great Britain)
Mr. J.V.S. Wilkinson	(British Museum)[4]

Empire Scientific Conference

The National Archives hold some interesting papers concerning the Empire Scientific Conference. For example:

The Royal Society
Burlington House
London
13th December 1945

Dear Sir,

The Royal Society has been invited by His Majesty's Government to organise an Empire Scientific Conference in 1946. Arrangements had been made for this conference to open on 17th June and to continue until 8th July. The main purpose of the conference is to stimulate the interchange of scientific information and to encourage the cooperative spirit in scientific research within the Empire. To this end it is proposed that the topics chosen for the discussion shall be presented in the form of papers, and to be followed by discussions on the subject.

In this connection the Policy Committee of The Royal Society Empire Scientific Conference has directed me to enquire whether you would consent to serve as a member of the working party, under the chairmanship of Sir Richard Gregory FRS, which is to prepare a paper on the dissemination of scientific news to the public for discussions at the Conference. The following are some broad suggestions for initial discussions:

a)   Broadcasts, television and gramophone records.
b)   Films.
c)   Press and publishing.

For your guidance the following are being asked are serve on this Committee:

Sir Richard Gregory FRS	Chairman
Mr. Vincent Alford	BBC talks department
Professor E.N. da C. Andrade FRS	
Mr O.F. Brown	DSIR
Mr. Richard Calder	
Mr A.C. Chalkley	Temple Press
Mr. J.C. Crowther	
Professor Lancelot Hogben FRS	
Sir Julian Huxley FRS	
Mr. R.J.L. Kingsford	Cambridge University Press
Mr. H. Martin	Press Association

Should you accept this invitation a further communication will be sent you as to the time and place of the first meeting

I need hardly say that your consent to undertake this work on behalf and the Society will be greatly appreciated.

Yours faithfully,
John D. Griffiths Davis
Assistant Secretary

Later in the year on 15 March 1945, the Empire Scientific Policy Committee met. Those present were:

Sir Alfred Egerton FRS   Chairman – Royal Society
Sir Edward Appleton FRS
Sir David Chadwick
Sir John Fryer FRS
Professor A.V. Hill FRS
Sir Edward Mellanby FRS
Mr. J.D. Griffith Davis also attended.

1. Having received views in regard to the conference from Canada, Australia, New Zealand and South Africa, it was agreed that, while it was hoped that the conference would be held as soon as possible after the conclusion of the German War, the decision as to the actual date should be deferred until September 1945. In this connection the representatives in London of the Dominions, India, and the Colonies were to be informed that the conference would not take place this year.

2. It was agreed that it would be a great help of representatives of the Dominions and India could arrange to be in England in the autumn of 1945 to discuss, with the Policy Committee, details of the organisation of the conference.

3. Sir David Chadwick reported on the proceedings at the two meetings of the Working Committee.

4. It was agreed that the objective of the conference should be defined tentatively as follows:
   'To consider practical measures for promoting closer collaboration between scientific people and organisations in the countries of the British Commonwealth and Empire and to make recommendations'

5. The members of the Committee undertook to:
   a) Prepare statements as to the framing of the program of subjects to be discussed at the conference.
   b) To give further attention to the question have recommending names of persons who might be included in the delegations.[5]

These four Committees demonstrate the ethos of the Royal Society, which has remained unchanged since its foundation in 1662.

Hopefully these few pages demonstrate that in times of adversity the Fellows and the corporate body that is the Royal Society are available to help meet any and all engineering, medical and scientific challenges. Of course the people making up the Society will change with time as will the specialism each Fellow is known for. Whilst the Royal Society continues to strive for academic excellence in its new recruits, one can imagine a willing pool of world experts available for every challenge presented to it.

Appendix I[1]

Members of the various scientific committees

21st August 1942
Dear Lucker,

As I explained over the telephone, Corell Barnes had passed your letter of the 20th August to me as Joint Secretary of the Scientific and Engineering Advisory Committees. I am able to answer a number of your questions out of hand with the information I have here or have been able to secure during the day. I hope to be able to deal with the more searching questions when I have had replies from the Department's concerned. Since I cannot use Sir Stafford's name these replies are not to be expected without some delay particularly as regards the research establishments and institutes.

Question 1

Scientific Advisory Committee
President of the Board of Education Chairman
Sir Henry Dale FRS
Professor A.V. Hill FRS
Sir Alfred Egerton FRS
Sir Edward Appleton FRS   Secretary DSIR
Sir Edward Mellanby FRS   Secretary Medical Research Council
Mr W.W.C. Topley FRS   Agricultural Research Council
Secretaries   Mr D.H.F. Rickett and Mr. H. Everett,
                                 War Cabinet Offices

Engineering Advisory Council
President of the Board of Education Chairman
Rt. Hon. Viscount Falmouth
Mr J.R. Beard
Dr A.V.M. Fleming

Mr W.T. Halcrow
Professor B.W. Holman
Dr C.C. Peterson
Sir Harry Ricardo FRS
Mr Andrew Robertson FRS
Secretaries   Mr H. Everett, War Cabinet Offices,
                                     Mr H. Wooldridge DSIR, Private
                                     Secretary to Sir Edward Appleton FRS

Question 2

Aeronautical Research Committee
Independent: Sir Henry Tizard FRS
                Sir Leonard Bairstow FRS
                Professor W.J. Duncan FRS
                Professor G.T.R. Hill
                Sir Bennett M. Jones FRS
                Sir Richard V. Southwell FRS
                Sir Geoffrey Taylor FRS
Official: Sir Edward Appleton FRS
                Sir Charles Darwin FRS
                Sir William S. Farren FRS
                Dr H.J. Gough FRS
                Dr N.K. Johnson
                Sir David R. Pye FRS
                Mr C.S. Wright

Admiralty Advisory Panel on Scientific Research and Development
Mr C.S. Wright Chairman
Sir Edward Appleton FRS
Sir Ralph Fowler FRS
Captain Wiley RN
Captain Hitchins RN

Ministry of Supply Advisory Council on Scientific Research and Technical Development
Sir Frank E. Smith FRS   Chairman
Sir Edward Appleton FRS
Sir William Bragg FRS
Lt-General Laurence Carr
Sir Charles D. Ellis FRS
Sir L.F. Freeman

Professor W.E. Gardner
Sir Ian M. Heilbron FRS
Professor A.V. Hill FRS
Sir William Larke
Sir John E. Lennard-Jones FRS
Sir David Pye FRS
Sir Robert Robinson FRS
Sir Robert Robertson FRS
Mr J. Rogers   Chairman Ammunition Board
Sir Richard Southwell FRS
Dr R.E. Stradling FRS (Ministry of Home Security)
Professor J.I. Taylor
Sir Henry Tizard FRS
Mr C.S. Wright   Admiralty

Ministry of Supply Committee on Armaments Development

Sir Henry L. Guy FRS	Chairman
Engineer Vice-Admiral Sir J. Preece	Admiralty
Captain C. Beaver	Director of Naval Ordnance, Admiralty
Major-General B.D. Hickes	Director of Staff Duties (Weapons), War Office
Dr H.J. Goff	Ministry of Supply
Major-General E.M.C. Clarke	Director of Artillery, Ministry of Supply
Mr A.F. Dobbie Bateman	Ministry of Supply
Air Commodore G.A.H. Pidcock	Director of Armaments Development, Air Ministry and MAP
Air Commodore P. Huskinson	President, Air Armaments Board, Air Ministry and MAP
Sir Edward Appleton FRS	

Accessors: Mr Ralph Freeman
              Mr W.T. Griffiths
              Mr H. Hillier
              Sir John E. Lennard-Jones FRS
              Mr W. Morgraph
              Sir Clifford Paterson FRS
              Sir Robert Robertson FRS

Ministry of Home Security Civil Defence Research Committee

Sir Edward Appleton FRS	Chairman
Dr E.F. Armstrong FRS	

Lord Baker FRS
Professor J.D. Bernal FRS
Sir Charles Darwin FRS
Sir Ralph Fowler FRS
Professor A.J.S. Pippard FRS
Sir Richard Southwell FRS
Sir Geoffrey Taylor FRS
Professor W.N. Thomas
Accessors: Mr C.S. Wright   Admiralty
            Sir David Pye FRS   MAP
            Professor H.J. Coff Ministry of Supply

Question 3

Ministry of Supply
Mr J. Davidson Pratt OBE   Controller of Chemical Defence Development
Dr E.T. Paris
Dr F. Roffey   Controller of Medical Research
Mr W.A. Robotham   Chief Engineer, Tank Designer

Ministry of Home Security Building Technical Advisory Committee
Mr David Anderson   Chairman
Mr P. Good   Chairman, Joint Lighting Committee
Mr H.T. Young   Chairman, Electrical Advisory ARP
                            Committee
Sir Reginald E. Stradling FRS Camouflage Committee

Medical Research Committee
Lord Balfour of Burleigh
Mr W. Goodenough
Mr J.G. Stuart MP
Professor G.E. Gask
Mr L.G. Witts
Sir Charles R. Harington FRS
Dr W.W.C. Topley FRS
Mr A.J. Clark FRS
Professor T.R. Elliott FRS
Lord Adrian FRS
Sir W. Wilson Jameson
Sir Edward Mellanby FRS   Secretary

National Institute for Medical Research
Sir Henry Dale FRS   Director

Industrial Health Research Board
Sir W. Wilson Jameson   Chairman

Agricultural Research Council
Sir Thomas Middleton FRS   Chairman
Sir Joseph Barcroft FRS
Sir Marrik Burrell CBE
Professor E.P. Cathcart FRS
Sir William Cecil Dampier FRS
Robert Blyth Greig
Professor T.J. Mackie
Spencer W. Mount
Sir Robert Robertson FRS
Sir Edward Salisbury FRS
Mr John Smith
Professor Sir William Wright Smith FRS
Professor W.W.C. Topley FRS   Secretary
Professor D.M.S Watson FRS

Agricultural Research Institutes
Number 26 and the names of their directories can be ascertained if necessary

Question 4

Principal Officers of the DSIR Advisory Council
Lord Riverdale of Sheffield
Sir Joseph Barcroft FRS
Mr G.M.B. Dobson FRS
Viscount Falmouth
Lt-Col Sir John Greenly
Sir Harold B. Hartley FRS
Professor A.V. Hill FRS
Sir Felix Pole
Sir Robert Robinson FRS
Sir Frank E. Smith FRS
Mr S.K. Thornley
Sir Edward Appleton FRS   Secretary
Mr L. S. Lloyd
Mr A.L. Hetherington

Mr G.R.D. Hart
Mr W.A.M. Murray

National Physical Laboratory
Sir Charles Darwin FRS

Fuel Research Station
Mr F.S. Sinnatt FRS

Low Temperature Research Station
Mr F. Kidd FRS

Ditton Research Laboratory
Mr C. West

Torry Research Laboratory
Mr G.A. Reay

Building Research Station
Mr J.C. Evans

Forest Products Research Laboratory
Mr W.A. Robertson

Road Research Laboratory
Sir William H. Glanville FRS

Chemical Research Laboratory
Mr B.D. Pratt

Geological Survey and Museum
Sir Edward B. Bailey FRS

Water Pollution Research Laboratory
Mr A. Parker

Pest Infestation Research Laboratory
Mr G.V.B. Herford

Signed Mr H. Everett

15th September 1942
Dear Lucker,

After, I fear a considerable delay, I am now able to give you further details of the matters referred to in your letter to Gorell Barnes of the 20th August. I have had some difficulty in securing this information, but I think the picture is now complete at least as far as question 3 is concerned. The result is as follows:

Admiralty **Research Establishments**	**Director** (or senior civilian scientist[2])
Admiralty Research Laboratory	Colonel A.V. Kerrison
Admiralty Signal Establishment	G. Shearing (communications) C.E. Horton (R.D.F.)
Anti-Submarine Experimental Establishment	B.S. Smith
Mine Design Department	A.B. Wood
Royal Naval Torpedo Factory	G.W. Austin
Admiralty Compass Observatory	W.G. Heatly
Engineering Department, Portsmouth Dockyard	I.G. Slater
Admiralty Engineering Laboratory	G.B. Fox
Bragg Laboratory, Sheffield	F.G. Barker

University Teams working on Radio Valve Research under D.S.R., Admiralty

Clarendon Laboratory, Oxford	T.C. Keely
Birmingham University	Sir Marcus Oliphant FRS

Ministry of Supply **Research Establishments**	**Chief Officer**
Air Defence Research and Development Establishment, Pale Manor Farm, Great Malvern	Professor J.D. Cockroft FRS
Chemical Defence Experimental Station, Porton, Near Salisbury, Wilts	Air Commodore G. Combe RAF

Chemical Defence Research Establishment, Sutton Oak, St. Helens, Lancs	J.W.C. Phillips PhD
Department of Tank Design, Wood Lee, Egham, Surry	A.A.M. Durrant
M.D.1, Whitchurch, Nr Aylesbury, Bucks	Brigadier M.R. Jefferis
Norfolk Flax Establishment, Flitcham Abbey, King's Lynn, Norfolk	G.O. Searle
Ordnance Board (External Ballistics Section), Chislehurst, Kent	Colonel A.H.D. Phillips
Projectile Development Establishment, Cilwendig Park, Boncath, Pembrokeshire	W. Blackman
Research Department, Shrewsbury, Salop	Brigadier J.L.P. Macnair
Signals Research and Development Establishment, Warnham Court, Nr Horsham Sussex	Sir John D. Cockroft FRS
Armaments Design Department, The Grange, Knockholt, Kent	Captain C.T. Nuthall
Coast and Anti-Aircraft Defence Experimental Establishment, Albert Drive, Llandudno Junction, Caernarvons	Lieutenant-Colonel C.L. Ferard
Experimental Bridging Establishment, Christchurch, Hants	Major G.M. Hunt
Experimental Demolition Establishment, Christchurch, Hants	Major J.A. Hill
Experimental Establishment, Pendine, Carmarthen	Captain S.A. Pears
Experimental Establishment, Shoeburyness, Essex	Colonel H.S. Lickman

Experimental Tunnelling Establishment, Homelands, Christchurch, Hants	Lieutenant-Colonel C.D.A. Fenwick
Wheeled Vehicles Experimental Establishment, Farnborough, Hants	Major E.C.A. Wood

Ministry of Aircraft Production

Experimental Establishments	**Chief Officer**
Royal Aircraft Establishment, South Farnborough, Hants	Sir William S. Farren FRS
The Aeroplane and Armament Experimental	Air Commodore R.H. Mansell
Establishment, Royal Air Force, Boscome	Mr E.J. Jones (Chief Technical
Down, Amesbury, Wilts	Officer)
Marine Aircraft Experimental	Group Captain P.A. Morton
Establishment, Royal Air Force, Helensburgh,	Mr H. Garner (Chief Technical
Dumbartonshire	Officer)
TRE, Great Malvern, Worcester	Mr A.P. Rowe
Royal Air Force Station, Defford, Worcester	Group Captain P. King
Torpedo Development Unit, Royal Air	Wing Commander R.E. Burns
Force, Gosport, Hants	Mr F.H. Beer (Senior Technical Officer)
Gunnery Research Unit, Royal Air Force, Exeter, Devon	Wing Commander H. Ford Sir Bennett Jones FRS (Scientific Officer in Charge)
Airborne Forces Experimental Establishment	Group Captain L.G. Harvey Squadron Leader N.H. Sharpe Royal Air Force Station Ringway, (Ringway)

Manchester	Mr G. Jennings (Chief Technical Officer)
Orfordness Research Laboratory, Orford Ness, Woodbridge, Suffolk	W.E. Candler
Bombing Trials Unit, Royal Air Force Station, West Freugh, Stranraer	Wing Commander A.H. Green

Ministry of Agriculture and Fisheries
Agricultural Research Institutes in Great Britain

Institutes	**Director or Acting Director**
Veterinary Laboratory Weybridge, Surrey	Professor T. Dalling
Foot and Mouth Disease Research Station	Dr I.A. Galloway
Grass Improvement Station Dodwell-Drayton, Warwickshire	Sir R. George Stapledon FRS
Plant Pathological Laboratory Harpenden	Sir John C.F. Fryer FRS
Institute for Research in Agricultural Engineering, Askham Bryan, Yorkshire	S.J. Wright

Under direct control of Agricultural Research Council

Field Experimental Station Compton, Near Newbury, Berks	W.S. Gordon
Unit of Animal Physiology, Cambridge	Sir Joseph Barcroft FRS
Unit of Soil Enzyme Chemistry, Rothamsted, Herts	Juda H. Quastel FRS

Agricultural Research Institutes financed wholly or in part by maintenance grants made by the Department of Agriculture and Fisheries, or by the Department of Agriculture for Scotland, and under the general co-ordination of the Agricultural Research Council in association with the Ministry or the Department

England and Wales

Soils and other subjects
Rothamsted Experimental Station,
Harpenden,   Sir E. John Russell FRS
Herts

Fruit and Horticulture
Long Ashton Research Station, Bristol   Professor B.T.P. Barker

East Malling Research Station, Kent   R.G. Matton

Institutes
Horticulture
Horticultural Research Station, Cambridge   D. Boyes

Glasshouse Crops
Experimental and Research, Cheshunt, Herts   W.F. Bewley

Plant Breeding and Genetics
Plant Breeding Institute, Cambridge   H. Hunter
Welsh Plant Breeding Station, Aberystwyth   T.J. Jenkin
John Innes Horticultural Institute, (Fruit dept),
Merton Park, S.W. 19   Dr C.D. Darling

Plant Physiology
Research Institute in Plant Physiology   Professor   V.H.   Blackman
                                              FRS
Imperial College, 17 Berkeley Place, S.W. 19

Animal Diseases
Institute of Animal Pathology, Cambridge   Professor T. Dalling
Research Institute in Animal Pathology, Royal
Veterinary College, Streatley, Berks   Professor T.J. Bosworth

Animal Nutrition
Animal Nutrition Institute, Cambridge
Dairying   Sir John Hammond FRS
National Institute for Research in Dairying,
Shinfield, Reading   Professor M.D. Kay

Institute of Agricultural
Parasitology, St. Albans   R.T. Leiper FRS

Agricultural Economics
Agricultural Economics Research Institute,
Oxford   C.S. Orwin

Scotland

Soils
Macaulay Institute for Soil Research,
Aberdeen   W.G. Ogg

Plant Breeding
Scottish Society for Research in Plant
Breeding, Corstorphine, Edinburgh   W. Robb

Animal Diseases
Animal Diseases Research Association,
Moredun   J. Russell Greig

Edinburgh

Animal Nutrition
Rowett Research Institute, Aberdeen   Lord John Boyd-Orr FRS

Animal Genetics
Institute of Animal Genetics, Edinburgh   A. Greenwood

Dairying

Hannah Dairy Research Institute, Kirkhill   Norman C. Wright

Ayrshire

NB during the War, the Agricultural Research Council also assumed the main financial responsibility for the Bureau of Animal Population, Oxford, for work on rodent control.

Medical Research Council

Research Establishments	**Director**

a) Separate Establishments wholly maintained by the Council.

National Institute for Medical Research (Hampstead and Mill Hill)	Sir Henry Dale FRS
Dunn Nutritional Laboratory, Cambridge	Dr L.J. Harris

b) Establishments maintained jointly by the Council and by the Institutions within which they are placed.

Department of Clinical Research, University College Hospital Medical School, London	Sir Thomas Lewis FRS
Clinical Research Unit, Guy's Hospital, London	Dr R.T. Grant FRS
Neurological Research Unit, National Hospital for Nervous Diseases, Queen Square, London	Dr K.A. Carmichael
Physiological Laboratory, Army Fighting Vehicles School, Lulworth	Dr O. M. Solandt

c) Establishments wholly maintained by the Council but located within other Institutions.

National Collection of Type Cultures of Micro-Organisms (Lister Institute, Elstree)	Curator: Dr R. St. John-Brooks

Standards Laboratory for Serological
Reagents

In charge: Lieutenant-
Colonel (School of
Pathology, Oxford) R.A.F.
Bridges

Radiotherapeutic Research Unit
(Hammersmith Hospital, London)

Dr Constance Wood

Department of Research in Industrial
Medicine (London School of Hygiene and
Tropical Medicine)

Professor A.W. Ellis

Appendix II[1]

Minutes of the Scientific Advisory Committee

War Cabinet
Scientific Advisory Committee
Minutes of the first meeting of the Committee

Held at Richmond Terrace on Thursday, 10th October
1940 at 3:30pm

Present:
The Right Hon. Lord Hankey FRS, Chancellor of the Duchy of Lancaster
– Chairman
Sir William Bragg FRS
Professor A.V. Hill FRS
Sir Alfred Egerton FRS
Sir Edward Appleton FRS
Sir Edward Mellanby FRS
Sir Edwin Butler FRS

The following were also present:
The Viscount Falmouth
Group Captain W. Elliot joint Secretary to the Committee
Professor W.W.C. Topley joint Secretary to the Committee

1. **Opening statements by the Chairman.**
   The Chairman made an opening statement dealing with the origin of the
   Committee and various points concerned with its scope and its relations
   to other scientific activities of the Government (attached as an appendix).

2. **Completion of a survey of scientific activities in, or supported by, government departments**

Agreed:

a)   That it would be of considerable value to the Committee to have as a background for their future work a survey of the various scientific activities which were being conducted in Government Departments and in Establishments receiving Government grants.

b)   As a first step to invite:

Sir Edward Mellanby FRS, Dr E.V. Appleton FRS and Sir Edwin Butler FRS to submit statements dealing with the scientific activities coming within their respective purviews, showing in each case:

   i)   The title or character of the work
   ii)   The individual or members engaged in the work
   iii) In the case where work was being done by a Committee, its terms of reference.

c)   That the secretaries will collect the statements in b) above for submission to the Committee.

d)   To leave to the discretion of the Chairman the question of an approach on similar lines to the defence departments and other government departments.

3. **Constitution of the Committee**

Professor Hill mentioned that a suggestion had been submitted by Sir Richard Gregory that a representative of the British Association should be appointed to the Committee.

The Committee:

a)   Recalled that it was not within the terms of its constitution to increase its membership, nor did it consider any increase desirable.

b)   Accordingly invited Professor Hill to confirm to Sir Richard Gregory that his proposal could not be accepted.

4. **Letters from inventors**

It was agreed:

a)   That it was not part of the work of the Committee that it should be a Board of Inventions or a Court of Appeal for the claims of inventors.

b)   That all letters from inventors should be sent to Dr Appleton, who would assume responsibility for acknowledging them and for sending them to the appropriate department for consideration.

5. Application for employment by the Committee

It was agreed:

That members of the Committee receiving applications for employment should reply to them individually, explaining that the services of the applicants could not be accepted by the Committee, and reminding them of the existence of the Central Register kept by the Ministry of Labour.

6. Attendance of other scientists at meetings of the Committee

It was agreed:

a) That the Committee should have the right to invite the attendance of other scientists if they considered that their work would be facilitated thereby, each case being considered on its merits.

b) Arising out of this, that Professor Appleton should ascertain whether Dr Stradling would be willing to attend the next regular meeting of the Committee, to be held at 3.00 pm on Monday, 21 October, and, in that event, to invite him to be present.

7. Invitation to the Lord President of the Council

The Chairman was invited to extend a cordial invitation to the Lord President of the Council to meet the members of the Committee at a special meeting which would be called at 3:00 pm on Tuesday 15th October.

8. Interpretation of the terms of reference

Consideration was given to the interpretation of the Committee's Terms of Reference as follows:

a) To advise the government on any scientific problems referred to it.

It was agreed:

a) That it would be the concern of the Committee to deal only with scientific problems referred to it by Government Departments, recognised scientific bodies, by all by scientists of repute; **not** by inventors (see conclusion 4(a) above).

b) To advise government departments, when required, on the selection of individuals for particular lines of scientific inquiry for the membership of Committees on which scientists are required.

The Chairman reported that he had agreed with the Minister of Labour on steps for avoiding any possible overlap with the Central Register.

It was agreed:

b) That this would not occur since the Committee would not be concerned with subordinates staff appointments but only with the appointment of scientific advisers on a higher level.

    c)   To bring to the notice of the government promising new scientific or technical developments which may be of importance to the war effort.

It was agreed:

    c)   That members of the Committee should turnover in their minds possible recommendations under this heading and submit them to the Committee at its next meeting – preferably in writing to the secretaries in advance.

    d)   That the Chairman would invite the Lord President of the Council to circularise all Government Departments calling attention to the activities of the Committee under (a) and (b) above.

9. Future procedure

The Committee agreed:

    a)   That a special meeting should, subject to Sir John Anderson's agreement, be held on Tuesday 15th October at 3:00 pm with the Lord President of the Council.

    b)   That the next regular meeting of the Committee should be held on Monday 21st October.

Annex

Opening statement by the Chairman at the first meeting of the Scientific Advisory Committee held on Thursday 10th October 1940

Origin of the Committee:

This Committee has been formed in response to a suggestion made twice, I think, by the Royal Society for closer cooperation between themselves, as representing the body of scientific opinion and experience in this country, and the government in the prosecution of War.

On the part of the Government the formation of the Committee represents a whole-hearted desire to obtain the utmost cooperation from the vast storehouse of brains, scientific knowledge, research and enterprise at the command of the Royal Society.

The formation of the Committee, besides achieving this objective, should, I think, strengthen the confidence of the public by providing a strong proof that science is being given its due place in the scheme of our national war effort. In fact the reception that the announcement has been given in the London and Provincial press already provides evidence that a favourable first impression has been created.

It will be our task to see that this favourable impression is maintained and strengthened, and, what is even more important, that useful results are obtained from our efforts.

Science already harnessed to a considerable extent
Science, of course, has already been harnessed to the war effort of the State to a very considerable extent – more widely, I suspect, than is generally realised except by those whose business it is to know, such as those present round this table.

Three research bodies
To begin with we have the solid nucleus of Government Research Institutions by represented round this table by the Secretaries of:

- The Department of Scientific and Industrial Research.
- The Medical Research Council.
- Agricultural Research Council.

With their respective research establishments, and their ramifications of organised research by individual scientific workers.

Co-ordination
Co-ordination is achieved between them by periodical meetings of their respective Directors and by day-to-day contacts on points of detail.

These organisations are at the present time carrying out a great deal of war-work on behalf of the three Fighting Services, as well as for the civilian side of our war effort as represented by other Government Departments.

Service Departments
In addition, the Service Departments have their respective organisations for specialist research.

The Admiralty
The Admiralty, for example, have a Director of Scientific Research, and a number of establishments where research is conducted in the various branches of naval warfare, such as naval construction, gunnery, mines, torpedoes, submarines, signals, wireless developments and so forth.

Example of the magnetic mine
The combating of the magnetic mine, into which I had some insight at the time, was a good example of how it works. Scientists outside the Government played a part in the development of the numerous methods that were evolved

for dealing with this menace, and the Royal Air Force co-operated effectively in one of these methods.

The War Office

The research work of the War Office, I think, has been concentrated in the Ministry of Supply, which has an efficient organisation under the direction of Dr Gough FRS, with wide ramifications. It does a great deal of work for all three Services.

The Central Directing Body, on which Lord Falmouth represents me, is presided over by Lord Cadman, and the three services are represented thereon, as well as on the numerous Committees to which the detailed work is delegated. Science outside the services is also well represented on this Committee.

In addition, the Ministry of Supply controls the long established Ordnance Board, the experimental work at Woolwich, the Chemical Warfare Organisation, including Porton, a wireless research station at Christchurch, and probably other research establishments that have not come particularly to my notice.

The Minister of Aircraft Production

The Minister of Aircraft Production has a Director of Research, under whom I notice in the official list has Directors of Research on a number of technical subjects, such as aircraft, aircraft engines, armament, equipment, etc. It also has its own Aeronautical Research Committee. But I am not familiar with the detailed organisation of research in this comparatively new Department, and I do not know how far outside scientists may be employed.

The Minister of Aircraft Production also has at his elbow Sir Frank Smith and Sir George Lee, who have control over a vitally secret branch of research which is of supreme importance to the Royal Air Force. They have an experimental establishment at Swanage.

Prime Minister's provisos

Before leaving the Services I should mention that the Prime Minister, in approving the establishment of this Committee, has laid down two provisos:

a) That the secrets upon which the various departments are now working (some of which he specifies) shall not be imparted to a new large circle.

b) That the time of the scientists and the Committees who are at present engaged in working for the Government shall not be unduly consumed.

'As I understand it', he observes, 'we are to have an additional support from the outside rather than an incursion into our interior'.

I do not think that we shall encounter any difficulty in adjusting our business to these conditions. This Committee is so constituted that some of its members, both inside and outside the Government circle, have already a knowledge and have rendered assistance in some of the matters of especial secrecy, and, if we find it necessary to approach those problems we can delegate them to those with the appropriate experience.

Civilian Departments' Research
The Civilian Departments of Government also conduct researches a over a wide field, and many distinguished scientists, both inside and outside the government, are assisting in this work.

For example, the Ministry of Home Security is conducting research into a variety of problems connected with its work (Sir Reginald Stradling FRS – Chief Technical Advisor).

The Cabinet Office, Home Office, Post Office, the Board of Trade in its subordinate departments, the Mines Department and Petroleum Department, the Colonial Office and Ministry of Food (Sir William Bragg and Sir Edward Mellanby and members of the Advisory Committee on Nutrition), among other departments have scientific bodies extending over a considerable field in the aggregate. And the Ministry of Labour has an Advisory Council with various Committees connection with the scientific side of its Central Register, which we shall have to bear in mind in connected with the second item of our terms of reference. Several of our members are associated with this work, and I anticipate no difficulty in achieving the necessary coordination.

Suggested compilation of Government research activities
Some of these activities have sprung into existence to meet the needs of the War, and I am not sure whether anyone has a complete knowledge of all of them. It has occurred to me that it might be useful if we made it part of our business to compile some kind of a list of such bodies, though shall have to be careful about some of the more secret Committees.

The Lord President of the Council
Another point, however, which I should like to mention is that I had a conversation yesterday on the subject of our work with the Lord President of the Council, whose predecessor appointed us, and to whom we have to report. Sir John Anderson had a scientific education and has kept himself abreast of scientific development as far as his busy life permitted, and, as I anticipated, I found that he takes a deep interest in this Committee. He asked me to say that he would very warmly welcome an opportunity to meet the members of the Committee and to discuss with them the work on which we are embarking today.

In conclusion I should like to say how much I appreciate the honour of presiding over so distinguished a Committee. I hope that we may together manage to make a real contribution to our war effort, and, in addition, possibly to produce results of lasting importance in strengthening and tautening up the links which already exist between the world of science and the Government and administration.

Appendix III[1]

Government Research Establishments

Explanatory statement on the diagram illustrating the organisation of research for defence purposes in Great Britain

Organisation

As indicated in the diagram, the ultimate responsibility for research to meet Defence needs rests upon the Ministers responsible for the Defence Departments and the Minister for the Co-ordination of Defence (Lord Chatfield). The work is carried out under the Directors of Scientific Research of the Admiralty, Air Ministry, Ministry of Supply (which is responsible for research and development for the War Office and whose program also includes items of work by the Admiralty and Air Ministry) and the Chief Research Advisor of the Ministry of Home Security. It is mainly conducted in Service research establishments, although, where expedient, arrangements are made for investigations in other laboratories. These include Government laboratories, such as those of the Department of Scientific and Industrial Research, industrial laboratories, such as those of the Research Associations (see below) and of private firms, and university laboratories.

The Directors of Research, the Chief Research Advisor to Ministry of Home Security and their staff consult each other fully and continuously on day by day developments. There are also numerous inter-Service Scientific and Technical Committees appointed to deal with various subjects. This close collaboration has led to the centralisation of investigations of common interest to two or more Services in the most appropriate service research establishment. The research programs of the three Service Departments (Admiralty, Ministry of Supply and Air Ministry) and of the Department of Scientific and Industrial Research are discussed at meetings of the Directors of Scientific Research under the Chairmanship of the Secretary of the Department of Scientific and Industrial Research held periodically and at least once a year.

There is, in addition, especially close contact between the Chief Research Advisor, in the Ministry of Home Security (who is concerned with ARP work) and the Department of Scientific and Industrial Research, since a good deal of work is carried out on his behalf and under his close supervision in the establishments of the Department.

It should be observed that the Department of Scientific and Industrial Research, the Medical Research Council, and the Agricultural Research Council are organisations for civil research under the control of the Lord President of the Council. They are thus not directly responsible for research for the Defence Services, though they may undertake specific items of work on repayment. The close contact between the Department of Scientific and Industrial Research and industry is illustrated in the diagram, while the Agricultural Research Council is very closely associated with the Ministry of Agriculture and Fisheries, the Department of Agriculture for Scotland and the Development Commission – the body recommending the provision of major part of funds for agricultural research.

The Medical Research Council also works in close association with Ministry of Health and indeed with all other Government Departments and, especially at the present time, with the Service Departments on all problems of health and disease requiring investigation.

Government Research Establishments
The following is a list of the research establishments under the control of the various departments indicated in the diagram.

Admiralty
Admiralty Research Laboratory
Admiralty Engineering Laboratory
Admiralty Compass Department
HMS Vernon
HM Signal School
HMS Osprey
RN Torpedo Factory
RN Cordite Factory
Naval Inspection Laboratory

Ministry of Supply (for the War Office)
Research Department, Woolwich
Chemical Defence Research Establishment
Air Defence Experimental Establishment
Signals Experimental Establishment
Projectile Development Establishment

Air Ministry
Royal Aircraft Establishment
Balloon Development Establishment
Radio Research Establishment
Aeroplane and Armament Experimental Establishment
Marine Aircraft Experimental Establishment

Ministry of Home Security
Various areas for full-scale field tests
Civil Defence Camouflage Research Establishment
Department of Scientific and Industrial Research
National Physical Laboratory
Departments for:

- Aeronautics.
- Electricity.
- Engineering.
- Metallurgy.
- Metrology.
- Physics.
- Radio.
- William Froude laboratory (tank research, design of ships hulls and propellers).

Fuel research station
Coal Survey of Laboratories (9 in number)
Food Investigation

- Low Temperature Research Station, Cambridge.
- Ditton and Research Laboratory (fruit).
- Torry Research Station (fish).

Pest Infestation Laboratory
Building Research Station
Forest Products Research Laboratory
Road Research Laboratory
Chemical Research Laboratory
Geological Survey and Museum
Medical Research Council
National Institute for Medical Research with departments for:

Bacteriology and Protistology, Biological Standards, Experimental Pathology, Pharmacology and Biochemistry, Physiology

National Collection of Type Cultures of Microorganisms (Lister Institute)
Standards Laboratory for Serological Materials (University of Oxford)
Dunn Nutritional Laboratory (University of Cambridge)
Department of Bacterial Chemistry (Middlesex Hospital)
Research Laboratory for Streptococcal Infections (Queen Charlotte's Hospital)
Department of Medical Statistics (London School of Hygiene)
Industrial Health Research Board

Agricultural research council
Field station at Compton, near Newbury, Berks

The Agricultural Research Council is also responsible for the scientific supervision of the research work carried out in the 26 autonomous, but state aided, research institutions, of which a list follows. These institutions cover the field of research relating to:

agricultural economics, agricultural engineering, animal breeding, dairying, glasshouse cultivation, grass land management, helminthology, horticulture, nutrition and disease, plant breeding, poultry, soils and plant physiology

Before the outbreak of War the Agricultural Research Council, in conjunction with the Departments of Agriculture, examined the staffs and research programs of all the research institutions to which reference is made below, with a view to ensuring that the staffs were well used to the best advantage in the event of War, whether in agricultural research service or in some other branch of technical service; that best use was made of existing knowledge in relation to food production, and that new problems likely to arise received attention. Since War broke out the position has been reviewed again. The Council are responsible for advising on the allocation of scientific workers included in the Agricultural Research Section of the Central Register of those having scientific, professional and technical qualifications.

State aided Agricultural Research Institutes
Agricultural and Horticultural Research Station, Long Ashton
Animal Diseases Research Association, Meredun Institute, Gilmerton
Animal Nutrition Research Institute, Cambridge University
Experimental and Research Station, Cheshunt, Herts
Hannah Dairy Research Institute, Kirkhill, Ayr

Horticultural Research Station, Cambridge University
Horticultural Research Station, East Malling
Institute for Research and Agricultural Engineering, University Oxford
Institute of Agricultural Parasitology (London School of Hygiene and Tropical Medicine)
Institute of Animal Genetics, Edinburgh University
Institute of Animal Pathology, Cambridge University
John Innes Horticultural Institution (fruit department)
Macaulay Institute for Soil Research, Craigiebuckler, Aberdeen
National institute for Research in Dairying, Reading University
National Institute of Agricultural Botany, Cambridge (Crop Improvement Section)
National Poultry Institute
Plant Breeding Institute, Cambridge University
Potato Virus Research Station, Cambridge University
Research Institute in Plant Physiology (Imperial College of Science and Technology)
Rothamsted Experimental Station
Rowlett Research Institute, Bucksburn, Aberdeen
Scottish Society for Research in Plant Breeding, Corstorphine, Edinburgh
Small Animal Breeding Research Institute, Cambridge University
Welsh Plant Breeding Station, University College of Wales, Aberystwyth

Industrial laboratories
The Research Associations referred to in the diagram are autonomous bodies formed on a national basis, within various industries. They are supported mainly by the industries concerned, but His Majesty's Government, through the Department of Scientific and Industrial Research, makes financial contributions. The following is a list of the Research Associations in receipt of grant aid:

Printing and Allied Trades Research Association
The British Association of Research for the Cocoa, Chocolates, Sugar Confectionery and Jam Trades
The British Boot, Shoe and Allied Trades Research Association
The British Cast Iron Research Association
The British Coal Utilisation Research Association
The British Cotton Industry Research Association
The British Electrical and Allied Industries Research Association
The British Electrical and Allied Industries Research Association
The British Food Manufacturers Research Association
The British Iron and Steel Federation (Industrial Research Council)

The British Launderers Research Association
The British Leather Manufacturer's Research Association
The British Non Ferrous Metals Research Association
The British Pottery Research Association
The British Refractories Research Association
The British Scientific Instrument Research Association
The Institute of the Automobile Engineers (Automobile Research Committee)
The Linen Industry Research Association
The Research Association of British Flour Millers
The Research Association of British Paint, Colour and Varnish Manufacturers
The Research Association of British Rubber Manufacturers
The Wool Industries Research Association

Particulars of the facilities, staff, and equipment available in the various laboratories of industrial firms and in university laboratories are included in a Register of Research Laboratories and Units compiled before the War by the Department of Scientific and Industrial Research. Particulars of the personnel available for research and development are included in the Registry of Individuals with Scientific, Technical and Professional Qualifications compiled by the Ministry of Labour and National Service.

Appendix IV[1]

Report from the Scientific Advisory Committee

**War Cabinet
Scientific Advisory Committee
28th February 1941**

Part 1

Composition and terms of reference
The Scientific Activities of the Various Government Departments
Scientific Activities of Department of Scientific and Industrial Research, the
Medical Research Council and the Agricultural Research Council
Scientific Activities of the Defence Services
Scientific Activities of Civil Departments Responsible for Research
Establishments
Scientific Activities of Departments not Responsible for Research
Establishments
The Scientific Sub-Committee on Food Policy
A Summary of Existing Scientific Organizations
Comments on the Survey of Scientific Activities
Problems Affecting Personnel

Part 2

Scientific cooperation between Great Britain, the Dominions, India and the
Colonies
Scientific cooperation between Great Britain, Canada and the United States
of America
The Enlistment of Help of Refugee Scientists
The Selection of Individuals for Particular Lines of Scientific Inquiry

Part 3

The future work of the Committee

Paragraphs annotated at the time, thought to be significant:

1. Whether it would be desirable to expand the summary contained in this report, or to compile any directory of scientific workers serving on Government Committees, or engaged in investigations on behalf of Government Departments, is a matter for future consideration on which we should welcome the views of Ministers and Government Departments.
2. The main conclusion that we should draw from our survey of scientific activities of the various Government Departments, is that they are far more extensive, and more effective, than is commonly realised. We are convinced that much of the criticism that has been offered on this score has been due to a lack of knowledge of the facts, itself resulting to some degree from the often unnecessary secrecy that obscures much of what is being done.
3. We think that we should keep this problem constantly before us, and, in particular, that every effort should be made to utilise the services of those relatively few persons of exceptional experience and ability for whom no place has yet been found; but, in view of all the difficulties, we believe that any general criticism of the present position is quite unjustified.
4. We believe that it would make for increased efficiency and economy if the organisation of every departmental scientific staff included the provision for regular consultation with a strong advisory council or committee of outside scientific experts, and that the facilities provided by the Department of Scientific and Industrial Research, The Medical Research Council and The Agricultural Research Council were utilised for the fullest possible extent.
5. This use of scientists should not stop at their employment on pure research, or on the development of apparatus or processes known to be required. The non scientific administrator cannot always recognise the problem as one likely to submit a scientific approach. There is need, therefore, for the closest collaboration between the scientific advisors and the scientific staffs of research establishments, on the one hand, and the operational administrative staffs on the other.
6. We believe that just as the defence services have reserves of officers and other ranks, who are called up in an emergency, so the scientific departments should in ordinary times make temperate appointments from the universities and other research institutions and should interest a wider range of scientist in their problems by freer interchange of personnel

between them and also other institutions. Such an interchange would obviously be facilitated by the adoption in all scientific departments of government, of the Federated Superannuation System for Universities.

7. The essential thing is to keep ahead of the enemy, and this depends on the speed with which all available resources are brought to bear on the particular problems at issue. If a justified fear of leakages is allowed to interfere with the greatest possible speed in finding a solution, the result may be much more disastrous than if some leakage had occurred, while in fact, the risk of leakage in wartime is, for a variety of reasons, less than in peacetime.

8. We have also been impressed by occasional failures to extract the utmost value for the results of such research during the stages of development and production. In view of the supreme urgency of speed, in a War that may depend in large part on the relative efficiency of the two belligerents at some critical phase, we would urge all possible steps should be taken to remedy any existing factors that make for delay.

9. It appears to us desirable, if practical, to simplify the existing procedure. Would it not be practicable, for example, to allow a member of the Services, subject to the approval of the officer commanding is unit, to forward an invention direct to the Government Department concerned and to receive a direct reply? We commend the subject to the Services for sympathetic consideration.

Appendix V[1]

Future of the Scientific Advisory Committee
21st November, 1945

Sir Henry Dale is coming to see the Lord President at 10:30am next Monday in order to discuss with him the future of the Scientific Advisory Committee.

Apart from those Ministers who have occupied the office of President, the Scientific Advisory Committee has, since its foundation, consisted of:

- The President and two Secretaries of the Royal Society.
- The three Secretaries of the Government Research Councils.

Sir Henry Dale and Professor A.V. Hill are retiring from their Royal Society offices on 30th November and are being succeeded by Sir Robert Robinson and Dr E.J. Salisbury respectively.

It is clear that membership of the Scientific Advisory Committee is ex officio and that in normal circumstances, therefore, Sir Henry Dale and Professor Hill had should now give way to their successors in office (it may be noted, however, that, at the intervention of the Lord President, Sir William Bragg remained a member of the Committee unto his death).

Sir Henry assumes, also, that the Chairmanship of the S.A.C. vests ex officio in the President of the Royal Society. This view has been contested by some of his colleagues on the Committee and I do not think that there is any written authority for it. Naturally, however, one would expect the President of the Royal Society to be Chairman of a Committee of this sort if he was a member.

As you will see from the attached note which he prepared for the Machinery of Government Committee two years ago, Sir Henry Dale does not approve of the present constitution of the Committee. His objections to it are threefold:

1. He does not think of the Royal Society in choosing a President should have to take into account the fitness of the candidate for this office to fill the entirely different office of Chairman of the S.A.C.

2. He feels that the official element in the Committee is too large. The Secretaries of the Research Councils already have the ear of the Lord President in their official capacity and they ought not to compose 50% of a body which is expected to give independent advice to the Government regarding the efficiency of its research organisations and to make recommendations regarding the scope of the Research Council's activities.
3. As to functions, he considers that the Committee should be used to a much greater extent than it has been in the past to advise the Government and scientific appointments (I know that he takes particular exception to the present system whereby the Lord President is not expected to consult any authoritative body of scientists about candidates for the DSIR Advisory Council and does not in fact consult anybody but the President of the Royal Society).

Since 1943 Sir Henry Dale's views have hardened. He tells me that he considers the Sir Henry Tizard's solution of the problem of securing authoritative scientific advice for the Government to be on the right lines. Under this, as you will remember, there would be a standing Committee with an independent scientist as Chairman, three or four other outsiders appointed after consultation with the Royal Society and the Institution of Civil Engineers and the three Secretaries other Research Councils. It's term of reference would be along the following lines:

i.   To advise the Government on the scientific implications of its general domestic and international policy insofar as the issues are not clearly the concern of a particular Government Department.
ii. To advise the Government on the scale and form of support to be given to scientific and technical education and to research work generally, and to keep under review the relation between the supply and demand for scientists and related pursuits.

It is fairly clear, however (see for instance the comment of flag 'A' below) that we cannot assume universal and automatic support for this form of set up among the scientists. In particular, I think that Sir Edward Appleton would want to consider his position very carefully before accepting it.

Sir Henry Dale is, as you know, fully aware of the proposed Committee on our Scientific Manpower and Resources and is most anxious that it should consider the position of the Scientific Advisory Committee and make recommendations to the Government about its future. I hope that you will agree that this is the right course, but if it is there is everything to be said for not 'revamping' the Scientific Advisory Committee until the new Committee has fully considered the position.

Fortunately, at the moment the Scientific Advisory Committee is not particularly active and no new question of major importance looks like coming before it in the next few weeks. Subject to your views, therefore, I have agreed with Sir Henry Dale that the best course to recommend to the Lord President is as follows:

a)   The existing SAC should not be reconstituted for the moment.

b)   The Lord President should see Sir Robert Robinson on his assumption of office (no doubt there will be plenty of opportunities for him to do this) and should explain to him that the future position of the SAC is to be considered by the Science Resources Committee and that we may hope for some recommendations from them on the subject in a few months' time; in the meantime he proposes that Sir Henry Dale should continue as Chairman of the Committee and Professor A.V. Hill should continue to be a member of it in order to clear up the few outstanding questions which are at present before it.

c)   The Lord President would, however, be very grateful if Sir Robert Robinson and Dr Salisbury could attend meetings of the Committee during this interim period so that, if the Committee receives a new lease of life in its existing form, they will be in a position to take over the places at present filled by Sir Henry Dale and Professor Hill.

Sir Henry Dale thinks that Sir Robert Robinson will agree to this course if it is put to him as a purely interim arrangement. My impression is that Sir Henry himself would be very pleased to continue as Chairman during the interim period.

I attach of your information a brief illustrative summary of some of the subjects which the Scientific Advisory Committee has considered since its inception.

M.T. Fleet

Appendix VI[1]

Meteorological research in the Royal Society

The formation of the Meteorological Research Committee by the Air Ministry, reported above, is an important and very welcome advance in the organisation of British Meteorological research. The Air Ministry Committee will at present confine its work chiefly to those problems which have an immediate practical application and are likely to be solved in a fairly short time. The Royal Society has been invited by the Air Ministry to cooperate with the Meteorological Research Committee by undertaking research on certain aspects of meteorology which, though of fundamental importance to the advance of the subject, may not have an immediate practical application.

The Council of the Royal Society have agreed to this request, and have entrusted the immediate responsibility for the work to the Gassiot Committee. This Committee, first constituted in 1871, was originally appointed to administer the Gassiot and other trust funds applicable to the maintenance of certain British meteorological and magnetic observatories, and to make recommendations as to their work. The terms of reference have now been enlarged to include the supervision of fundamental meteorological research such as has been asked for by the Air Ministry. The personnel for this Committee is appropriate for this purpose: the Chairman is Dr G.M. B. Dobson FRS, and the members are Sir Edward Appleton FRS, Sir David Brunt FRS, S. Chapman FRS, Sir Alfred Egerton FRS, Sir Henry Lyons FRS, Sir George C. Simpson FRS, Sir George Taylor FRS, Sir Gilbert Walker FRS, the Astronomer Royal, the President of the Royal Astronomical Society and the Director of the Meteorological Office.

These arrangements seem eminently suitable, because pure fundamental research which may have to be continued over a long period, and which requires the co-operation of many independent workers in university departments and laboratories, can probably be organised better by the Royal Society than by a Government Department, however much the latter may be in sympathy with the object of the researches. The Royal Society has already shown its willingness to further such work by appointing Dr E. Gluckauf to the Mackinnon Studentship, to enabled him to carry out meteorological research under the general guidance of the Gassiot Committee.

The general research in which the Air Ministry has sought the corporation of the Royal Society is that of radiation in the atmosphere. It is fairly generally accepted that the average temperature of the stratosphere – and of the warm region above it – is determined by an equilibrium between the energy absorbed, from both the downward solar radiation and the upward terrestrial radiation, and the energy radiated by the air in the stratosphere. Observations have shown that there are large variations of temperature in the stratosphere both with time and with place, but we know little as to why such variations occur. Few meteorologists, for instance, would care to be dogmatic us to why the stratosphere is some 50° C colder in equatorial regions than in polar regions.

The principal gases of the atmosphere are mostly very transparent to radiation, and the radiative conditions are largely governed by such a minor but polyatomic constituents as water vapour, carbon dioxide, ozone and possibly others. To understand the radiation equilibrium in the atmosphere it is necessary to know a) what gases are present which have important absorption bands for radiation: b) in what proportions they are present at different levels: c) how gases such as ozone are formed and decomposed; d) the adsorption co-efficients of these gases for different wavelengths, and whether the absorption and emission spectra have a continuous or fine line structure. Finally, when the above facts are determined, it will be necessary to calculate the radiation equilibrium temperatures at different levels and to determine how rapidly the temperature will change when the conditions alter.

The Gassiot Committee had been fortunate in enlisting expert help in particular branches of the subjects involved, and has formed three Sub-Committees to deal with different aspects of the above question. It is hoped in this way that the available relevant knowledge will be collected, and that those questions which require more experimental work may be defined, so that arrangements can be made for the work to be done, so far as this should prove possible under the present conditions.

Dr Dobson FRS is the Chairman of Sub-Committee A, which will deal with the chemical analysis and the observation of the amounts of polyatomic constituents of the air at different levels; other members of this Sub-Committee are Professors H. Dingle, F.A. Paneth FRS and Dr E. Gluckauf FRS.

Sub-Committee B will deal with the photophysics and photochemistry of the atmosphere, from the highest levels downwards, and, in particular, with the changing equilibrium of ozone and any other variable polyatomic constituents; the Chairman of this Sub-Committee is Sir Harrie Massey FRS, and members of are Professor K.G. Emeleus, Dr W.C. Price FRS and Dr J. Sayers.

Sub-Committee C will consider the radiation balance in the atmosphere; its Chairman is Professor S. Chapman FRS, and its members are Sir David Brunt FRS, Dr G.S. Callendar, Dr T.G. Cowling FRS, Dr A.R. Meetham, Sir George Simpson FRS and Sir Gordon Sutherland FRS.

There will be few who will not agree that the first meteorological problem thus selected for attack is a very important one and still fewer who will dispute its difficulty. It is hoped that the scientific effort which the Royal Society can bring to bear upon this work will lead to substantial progress.

The Gassiot Committee is aware that at laboratories and institutions in many parts of the world work has been undertaken that bears on these problems, and that the Committee would welcome co-operation in cases where the continuance of such work is possible in present circumstances.*

*Correspondence should be addressed to The Secretaries, The Royal Society, Burlington House and marked 'The Gassiot Committee', or to individual members of the Gassiot Committee at the same address.

Appendix VII[1]

Top Secret 'U'

The following is a list of Boards, Committees and Sub-Committees dealing with British Communication Security matters showing the Chairman, Secretary and the present G.C. & C.S. representatives.

Section A.

Organisation of the Cypher Policy Board which is a Cabinet Committee

Cypher Policy Board
Chairman:	Major General Sir Stewart Menzies	(Director General)
Secretary:	Captain (S) D.A. Wilson	(D.D.(C.S.A.))

Cypher Security Committee
Chairman:	Captain (S) D.A. Wilson	(D.D.(C.S.A.))
Secretary:	Commander G. Bull	(D.D.(C.S.A.)'s Section)
	Brigadier J.H. Tiltman (D.D.(4))	
	Commander (S) R. Dudley-Smith	(S.A.C.)
	Commander (S) K.L. Barrow (Supt. Production, Mansfield College, Oxford)	

Cypher Machine Development Committee
Chairman:	Air Commodore C.S. Cadell	(Air Ministry)
Secretary:	Commander T.R.W.B. Miller	(D.D.(C.S.A.)'s Section)
	Captain (S) D.A. Wilson	(D.D.(C.S.A.))
	Mr. P. Twinn	

War Office Cypher Machine Development Sub-Committee
Chairman:	Captain (S) D.A. Wilson	(D.D.(C.S.A.)
Secretary:	Mr. R.H. Wallace	
	Commander T.R.W.B. Miller	(D.D.(C.S.A.)'s Section)

Major W.G. Morgan
Mr. P. Twinn

Rockex II Sub-Committee
Chairman: Captain (S) D.A. Wilson   (D.D.(C.S.A.)
Secretary: Mr. R.H. Wallace
           Commander T.R.W.B. Miller   (D.D.(C.S.A.)'s Section)
           Mr. F.V. Freeborn
           Mr. A.M. Turing FRS
           Mr. H. Beil
           Mr. P. Twinn

Section B

Communication Boards and Committees responsible to the Chiefs of Staff's Committee

British Joint Communications Broad
Chairman: Group Captain Leonard Williams
Secretary: Commander W.E. Charles (deals mainly with British-U.S. Communications matters)

Sub-Committee "G" (Code & Cypher Sub-Committee)
The composition of this Committee is the same as the Cypher Security Committee, but in matters relating to British-U.S. Communications it reports to the B.J.C.B., not to the Cypher Policy Board.

Sub-Committee "C" (Methods & Procedures Sub-Committee)
Chairman: Group Captain Leonard Williams
Secretary: Commander W.E. Charles
           Lieutenant-Colonel L.E. Clark   (D.D.(C.S.A.)'s Section)

Radio Board
Secretary: Group Captain D.H. Johnson.

Speech Secrecy Panel
Chairman: Mr. A.J. Gill (G.P.O.)
Secretary: Mr. H.D. Bickley (G.P.O.)
           Commander T.R.W.B. Miller   (D.D.(C.S.A.)'s Section)

W/T Security Committee
(Reports to the Chiefs of Staff through the J.I.C.)
Chairman: Captain C.L. Firth   (Admiralty)
(Joint Secretaries):

Commander G. Bull	(D.D.(C.S.A.)'s Section)
Lt. Commander M. Davenport	(Admiralty)
Captain (S) D.A. Wilson	(D.D.(C.S.A.)
Commander (S) R. Dudley-Smith	(S.A.C.)

Section C.

Air Ministry Committees linked with but not responsible to any Inter-Service organisation

Air Ministry Cypher Machine Policy Committee
Chairman: Air Commodore C.S. Cadell
Secretary: Wing Commander E.C. Badcock
                Commander T.R.W.B. Miller   (D.D.(C.S.A.)'s Section)

Typex Development Committee
Chairman: Air Vice Marshal O.G. Lywood   (26th Group)
Secretary: S/Ldr. T.L. Walters
                Commander T.R.W.B. Miller   (D.D.(C.S.A.)'s Section)
                Mr. P. Twinn

17th August, 1945

Appendix VIII[1]

The Guy Report

1. Appointment of the Committee

I. The Committee was appointed in May 1943, by the Ministry of Supply with the concurrence of the Admiralty, Air Ministry, Ministry of Aircraft Production, Science Advisory Committee of the War Cabinet, and the War Office, with the following Constitution and Terms of Reference:-

Constitution

Chairman H.L. Guy FRS	
Members	
Engineering Vice-Admiral Sir George Preece	Admiralty
Rear-Admiral O. Bevir	Admiralty
Air-Commodore G.A.H. Pidcock	Air Ministry and MAP
Air-Commodore P. Huskinson (ret)	Air Ministry and MAP
Sir Edward Appleton FRS	Scientific Advisory Committee
H.J. Gough FRS	Ministry of Supply
Major-General E.M.C. Clarke	Ministry of Supply
A.F. Dobbie-Bateman	Ministry of Supply
Major-General L.D. Hickes	War Office
Assessors	
Ralph Freeman	
W.T. Griffiths	
H. Hillier	
Professor J.E. Lennard-Jones FRS	
W. Morpeth	
C.C. Paterson FRS	
Professor Sir Robert Robinson FRS	
Secretary J.W.L. Ivimy	Ministry of Supply

Terms of Reference

To review the machinery for the conduct of research, design and experimental work in connection with the development of guns, small arms, ammunition and cognate stores; to consider, in particular, the functions and organization of the Ordnance Board, the Armaments Design Department and the Research Department; and to make recommendations.

2. We have held 18 meetings, the first being 21st May 1942 and have also on 12 days visited the following establishments:-
Ordnance Board and External Ballistics Department, Chislehurst
Armament Design Department:-
> Headquarters, Gun Section, Check and Amendment Section, Knockholt
> Carriage and Ammunitions Sections, Halstead Place
> Small Arms Section, Cheshunt

Research Department
> Headquarters, Shrewsbury
> Internal Ballistics Branch, Cambridge and Woolwich
> Explosives Branch Sections, Woolwich and Swansea
> Metallurgical Branch Sections Woolwich, Swansea and Cardiff
> Proof and Experimental Establishment, Woolwich

Royal Ordnance Factories, Woolwich
Royal Aircraft Establishment, South Farnborough, Hants
Works of Messrs. Metropolitan-Vickers Ltd, Trafford Park
In the course of our visits to the Armaments Design Department and the Research Department we have held 82 interviews with members of the Service and civilian staffs representing typical cross-sections of the establishments.

3. We have had before us the following reports of former Committees touching the Armaments Development Organisation:-
> Report of Treasury Committee under the Chairmanship of Sir Arthur Duckham (1925)
> Report of Ministry of Supply Committee under the Chairmanship of Colonel J.J. Llewellin (1939)
> Report of a Joint Panel of the Scientific and Engineering Advisory Committees of the War Cabinet under the chairmanship of The Rt. Hon Lord Hankey FRS (1942)

4. We have heard evidence from the following:-
Dr J.W. Armit, Director-General of Explosives

Air-Vice-Marshal G.B.A. Baker, Senior Air Staff Offices, Coastal
  Command RAF
Engineer Vice-Admiral Sir Harold Brown, Senior Supply Officer
Group Captain L. De V. Chisman, Deputy Chief Superintendent of
  Armaments Design
Dr C.A. Clemmow, Superintendent of Ballistics Research
Air Commodore G. Combe, Vice-President Ordnance Board
Major-General D.S.C. Evans, Vice-President Ordnance Board
Professor W.E. Garner FRS, Leverhulme Professor of Physical Chemistry
  University of Bristol
C.D. Gibb, Director-General of Weapon and Instrument Production
Sir Stanley V. Goodall, Director of Naval Construction Admiralty
Dr R.H. Greaves, Director of Metallurgical Research
Colonel C.R. Hodgkinson, Deputy Chief Superintendent of Armaments
  Design
Brigadier O.F.G. Hogg, Director of Technical and Military
  Administration
Captain H.F. Howse RN, Proof and Experimental Officer
Dr J.W. Maccoll, Principal Scientific Officer External Ballistics
  Department
Sir Charles N. McLaren, Director-General of Ordnance Factories
G.S. McLay, Director-General of Gun Ammunition Production
Brigadier J.L.P. Macnair, Chief Superintendent Research Department
Captain C.T. Nuthall RN, Chief Superintendent of Armaments Design
Dr E.T. Paris, Controller of Physical Research and Signals Development
Colonel A.H.D. Phillips, Superintendent of External Ballistics
Dr H.J. Poole, Assistant Director of Scientific Research
Vice-Admiral A.F. Pridham, President Ordnance Board
Colonel G.O.C. Probert, Ordnance Board
Sir Robert Robertson FRS, Explosives Branch Research Department
Dr F. Roffey, Controller of Chemical Research
Dr G Rotter, Director of Explosives Research
Major-General F.B. Rowcroft, Director of Mechanical Engineers War
  Office
Rear-Admiral C.E.B. Simeon, Deputy Controller Admiralty
C.H. Stevens, Consultant to Armaments Design Department
Lieutenant-General R.M. Weeks, Deputy Chief of the Imperial General
  Staff, War Office

5. Our Assessors, appointed to assist us with specialised knowledge and
    wide experience in research and design outside the Government service,
    have heard the evidence and accompanied us on our visits. They have

recorded their conclusions in reports which have been of the greatest value to us in framing our recommendations. We desire to place on record our appreciation of the constant guidance they have given us despite the pressure of other commitments and our high estimation of the expert advice and judgement with which they have enlightened our deliberations. Their reports deal with a number of points which, though important, could not be considered by us in detail without unduly prolonging our session.

6.   We pay high tribute to the services of our secretary. The work of the Committee has been very exacting and Mr Ivimy has carried the main burden. We owe him a great debt.

II.   General

7.   In this report we confine ourselves to a short term plan directed to major objectives alone and calculated to produce immediate results. We recognise that much good work has been accomplished by the establishments concerned under very difficult conditions but are of the opinion that changes can and must be made now if these establishments are to be fully equipped to discharge the onerous and vital tasks which are theirs in the war effort.

8.   Our proposals involve the appointment of important new officials to occupy key posts vital to the working of the whole scheme. Our recommendations are therefore confined to broad issues, framing a structure the details of which can be modified and adjusted in accordance with the views of these new officials when appointed

9.   In drafting this Report we have taken the view that, in the middle of War, such changes only should be made, and in such a manner, as are possible without disorganization of current work. Any other course might involve dislocation, if not complete breakdown, of the machinery for Research and Design. Changes which although desirable are not imperative are not dealt with in this report and should be postponed until such times as they can be introduced without dislocation of current work.

10.   In making changes consequent on our proposals every endeavour must be made to carry the support and good will of the Service and civilian staffs at present in and operating the establishments affected. We recognise that the Service technical officer and user, and the professional engineer and scientist, alike have an important contribution to make and must be

suitably selected and properly used in the proper place in the organisation for armament development.

11. It is our intention that our proposals should fully recognise and maintain the position and authority of the ordering and approving authorities, the Director of Naval Ordnance (D.N.O.), the Director of Armament Development (D. Arm. D.), etc. and subject to the requirements of their programs, should establish the Head of the Research Department as the authority on armament research and responsible for the conduct of the work in his Department and similarly the head of the Armament Design Department as the authority on armament design and responsible for the conduct of the work of his department.

    The Ordnance Board has a wide and useful advisory function, but, in our opinion, should be executive only in so far as is necessary to stage and carry out trials. Its advice can be sought – when required – by the ordering authority, the Research Department, and by the Armaments Design Department. The Board also by the medium of its Proceedings disseminated knowledge of armament development between the three Services and other authorities. The independent position of the Board is a feature which should be maintained.

12. We accept as applicable to all the Services, a dictum of Marshal of the Royal Air Force, Lord Trenchard, quoted in evidence:-
    It is as much the duty of the technical departments to anticipate requirements as of the Air Staff to put them forward.

    The technical Directorates, the Research Department and the Armaments Design Department must, therefore, be motive forces in initiating and pursuing the development of weapons, and for that reason there should be at all stages the closest possible co-operation between the Directorates and Establishments and the Service Operational Staffs for the free exchange of information.

13. We have heard a considerable amount of evidence pointing to the desirability of the armament development establishments being in closer contact with the actual user in the Army to supplement the contribution of the military technical officers. We recommend that the Ministry of Supply and the War Office should in collaboration take such steps as they consider expedient to achieve this object.

14. In times of intense production effort the production services necessary to development are frequently ignored. The ultimate test of production must be the volume of suitable material which is produced and issued

to the armament Services. The experimental production requirements for a progressive armament development policy are very small in relation to main production programs, and should, and can under the direction of the chief production officers of the Ministries, take precedence over normal production.

In order to preserve proper balance within the Ministries on such matters, the heads of the Research and Armaments Design Departments should be able to deal on a basis of equality of status with the Director-Generals responsible for production.

15. As the establishments are under the Ministry of Supply we consider that the Controller-General of Research and Development (C.G.R.D.) should be responsible for their general administration, and for ensuring that they are adequately staffed and properly equipped to fulfil the functions required of them by all the Ministries concerned.

The heads of the Research and Armaments Design Departments and the President of the Ordnance Board (P.O.B.) should be directly responsible to C.G.R.D. that their respective departments are properly and efficiently organised.

The ordering authorities – D.N.O., D. Arm. D. etc., should continue to be responsible for formulating the programs of work and for the approval of the work when performed as satisfying the requirements of their respective Services.

We recommend that an Inter-Service Committee should be formed, consisting of the D.N.O., D. Arm. D., D. of A., the Head of the Research Department (C.S.A.R.) and the Head of the Armaments Design Department (C.E.A.D.) and P.O.B., under the Chairman chosen by them from one of themselves, to confer on armament research, design and development and in particular on matters of major technical and administrative policy concerning work for the three ordering authorities named carried out in the three departments.

It is expected that the Committee would be required to meet not more frequently than once a fortnight.

A Secretary of the grade of Principal should be appointed for the Committee's services, who would be located at the Ministry of Supply headquarters.

16 We recommend that serious consideration be given to establishing an engineering research laboratory to deal experimentally or analytically with those questions concerning the strength, endurance, stress distribution, and functioning of structural elements, etc., which are not at

present examined in any section of the Research or Armaments Design Departments.

The practice which has grown up of utilising the Engineering Department of the National Physical Laboratory for some of these purposes should be explored with the Department of Scientific and Industrial Research to decide the extent to which such arrangements can be made permanent.

We are informed that a similar need exists for this service for other activities of the Ministry of Supply.

If a suitable provision for this work cannot be made in this way, serious consideration should be given to establishing an engineering research laboratory within the department of the C.G.R.D.

17. We heard evidence of grave shortcomings in the present state of armament development which are directly traceable to failure to make adequate financial provision for research, design and development in the years preceding the War and to delays on the provision of necessary staff and facilities. Opportunities for research and experiment are greater in peace than in War. Once lost they can never be regained.

We consider it of vital importance to national security that in peace and War alike considerations of a financial or 'establishment' nature should not over-ride the primary importance of speed and continuity of technical development.

III Research Department

18. Having duly considered the facts and opinions brought before us, we make the following recommendations:-
    i)   That the title of the Department be altered to Armaments Research Department (A.R.D.).
    ii) That a scientist of high standing, suitable personality, and administrative ability, be appointed by the Minister of Supply, in consultation with the Admiralty and the Ministry of Aircraft Production, to direct and guide the work of the Department as Chief Superintendent of Armaments Research (C.S.A.R.). The scientific status of this office should be not less than that of Director of the National Physical Laboratory and its emoluments should be the same as those recommended in paragraph 19(i) for the Chief Engineer of Armaments Design.
    iii) That the Naval, Military and RAF personnel necessary to provide Service knowledge for the Department should form part of the staff of the C.S.A.R. Each Service should have its own senior representative

in the Department. All other Naval, Military and RAF personnel should be allocated to the sections, mixing and working with the staff of those sections and under the administrative control of the head of the section.

iv) That the necessary Naval, Military and RAF personnel should be nominated by D.N.O., D. of A. and D. Arm D. respectively in consultation with the C.S.A.R.

The number required in each main section will vary with the nature of the work and the character of the problems being investigated at any one time. A minimum number should be agreed between the C.S.A.R., D.N.O., D. of A. and D. Arm D. To be increased by agreement at any time for short or long term service as the occasion may require.

v) That the existing administrative branch of the Armaments Research Department be replaced by a new branch under a Secretary who should be directly responsible to C.S.A.R. for the administration of Headquarters and outstation branches of the establishment. The Secretary should be primarily an administrator having had a scientific or technical training and his post should carry a salary of £1,050 to £1,250 per annum, with corresponding status in the establishment. He should be responsible for providing efficient administrative service to the technical and scientific staff and relieving them as far as possible of routine correspondence and returns, while exercising no scientific or technical control over the work of the establishment. Each section or outstation should have its own administrative and clerical staff, posted to it by the Secretary and forming part of his Department, but responsible to the local Head of Section.

vi) That, in order to make the posts of Superintendents of the technical departments of the A.R.D. attractive to the best qualified and most experienced scientists in the country, these posts should be recognised as personal appointments and suitable holders should have salaries within the range of £1,400–£1,600. Similar changes in the status of other senior positions should be made as the occasion requires, and some additional appointments will be necessary.

vii) That excessive centralisation of scientific and administrative control of which there is some evidence should be relaxed in order to encourage initiative and enthusiasm for departmental heads and senior staff.

viii) That the C.S.A.R. should take such steps as seem to him necessary or desirable to foster enthusiasm and initiative, such as the institution of conferences of the senior Service and civilian members of his staff and by encouraging the free interchange of scientific thought and experience of different sections of the department itself.

ix) That in order to reduce loss of time and to relieve the O.B. it is desirable, whenever possible, for orders and instructions from D.N.O., D. of A. and D. Arm D., or other ordering authorities to be sent direct to A.R.D., with a copy to the O.B. as those ordering authorities may decide. A.R.D. should, within itself, co-ordinate these requests from the various Services.

x) That the Proof and Experimental Establishment be separated from A.R.D. to form a new technical establishment. A.R.D. should loan to the Proof and Experimental Establishment such civilian scientific staff as is necessary to it for the correct discharge of its functions.

xi) That the work on external ballistics research now conducted by the Ordnance Board be transferred to the A.R.D. who should post to the Board such sections and personnel as may be necessary for the best conduct of its work. Within A.R.D., work on ballistics should be organised into three branches dealing with internal, external and terminal ballistics, respectively.

xii) That the A.R.D. should avail itself fully of the resources of appropriate expert advice review and discussion offered by the machinery of the Scientific Advisory Council (S.A.C.). C.S.A.R. should closely coordinate the work of his Department with that of the Committees of the S.A.C.

xiii) That such part of the gun calculation section as is necessary to make the routine calculations to serve the Armaments Design Department shall be posted to and located in the appropriate Design section of A.D.D., although remaining a part of the Internal Ballistics branch of A.R.D.

xiv) That suitable members of A.R.D. staff be posted to A.D.D. to provide day to day advice in their fields eg explosives, metallurgy, pyrotechnics.

xv) That in view of the work done in the Explosives Branch of A.R.D. in developing manufacturing processes to the pilot plant stage a suitable staff of chemical engineers should be included in the establishment.

IV Armaments Design Department

19 We recommend:-
i) That a highly qualified and experienced Mechanical Engineer be appointed, by the Minister of Supply in consultation with the Admiralty and the Ministry of Aircraft Production, Chief Engineer and Superintendent of Armaments Design (C.E.A.D.) in complete charge of A.D.D. and given high status. The status and emoluments of his post should be identical with those of the post of Director of

Naval Construction. We realise that his means a salary of £2,500 a year at present, and in recommending it we are not necessarily recommending that rate as sufficient but as the highest existing rate for such a post in the Government service.

ii) That the introduction of C.E.A.D. be accompanied by the establishment of an engineering section of highly trained and experienced mechanical engineers headed by 4 to 6 Principal Design Engineers rated from £1,200 to £2,000, with supplementary lower categories of Design Engineers, to provide a team of about 40 in all, recruited from the best men available from inside or outside the Department. This staff will be additional to and superimposed on the existing organisation and will first concentrate on new work or special investigations now needed, and later infiltrate into the sections.

iii) That the existing administrative branch of the Armament Design Department be replaced by a new branch under a Secretary who should be directly responsible to C.E.A.D. for the administration of the Headquarters and outstation branches of the establishment. The Secretary should have had a technical training and his post should carry a salary of £850 to £1,050 per annum, with corresponding status in the establishment. He should be responsible for providing efficient administrative service to the technical staff relieving them as far as possible of routine correspondence and returns, while exercising no technical control over the work of the establishment. Each section or outstation should have its own administrative and clerical staff, posted to it by the Secretary and forming part of his Department, but responsible to the local Head of the Section.

iv) That in order to continue current work without interruption the existing Service and civilian organisation, with the exception of the post of Chief Superintendent of Armaments Design, be retained to carry on the work it now handles. An additional appointment of Deputy Chief Superintendent should be made on the nomination of the Admiralty. The three posts of D.C.S.A.D. will remain permanently and be filled by the senior representative of each Service in the establishment.

v) That the new engineering section will draw on existing staff for particular projects, thus assisting infiltration. Selection for special work in this way, if justified in the result, should form grounds for advancement in pay and on occasion in status.

vi) That in due time C.E.A.D. will either introduce new highly qualified designers into executive sections or confirm in or promote to such posts existing Service or civilian staff, the choice being made on grounds of suitability alone.

vii) That the selection of individuals for higher posts down to and including posts of Superintendent of Sections, who under C.E.A.D. are responsible for design in the sections, shall be a matter for the C.E.A.D. in consultation with the principal ordering and approving authorities concerned with each Section.

viii) That when as a result of action under (vi) above, a civilian engineer takes over a post previously occupied by a Service officer, C.E.A.D. will, in consultation with D.N.O., D. of A, or D. Arm D., arrange to put into the Section affected a Service officer or officers to supply Service knowledge and experience and to ensure that designs fulfil Service requirements.

ix) That for the latter purpose, some of the Service staff should have up-to-date user experience and therefore be frequently changed and drawn from various sources including the maintenance and gunnery services.

x) That in appropriate cases of special development recourse can with advantage be made to the use of the project engineer to follow up and co-ordinate a particular development through the design and development stages.

xi) That the emoluments of senior design posts such as those of Superintendent should be related to the post and any serving officer who may be appointed to such a post should receive the pay appropriate to the post and not Service pay.

xii) That as our enquiries and observations lead us to the conclusion that promotion primarily by seniority has had harmful effects on work or outlook of the design and draughtsman staff, the application of this principle to the detriment of promotion by merit be abandoned. In addition further means should be found for rewarding merit apart from promotions.

xiii) That to meet the difficulty in obtaining draughtsmen, to relieve the load on the A.D.D., and to make increased use of the experience of industry, design work be placed with industry by the A.D.D. to a greater extent than has been done in the past.

xiv) That the number of Senior Design Officers should be increased.

xv) That the existing experimental workshop facilities immediately available to the Department be increased.

V.   Ordnance Board

20. While we have had before us as witnesses the President and Vice-Presidents of the O.B. and have also received much evidence and testimony on the relations of the O.B. to the work of the Design and Research

Department, we have not yet sufficiently examined the constitution of the Board, its detailed work and method of operation, to be prepared to make recommendations for any fundamental change in the constitution of the Board, its functions and authority, and we therefore limit ourselves to such matters as may help to expedite the conduct of the Board's existing work. Certain other of our recommendations follow from changes we propose in A.R.D. and A.D.D. We desire, however, to emphasise that the functions and duties of the Board as laid down in 'Instructions for the Ordnance Board', dated 7.6.39, need clarification and revision.

21. Subject to the foregoing, we consider that the functions of the Board should be defined broadly as follows:-

   i)   The Ordnance Board's duties are primarily advisory. It advises the technical Armaments Directorates of the three Services on such armament matters as are referred to it by those Directorates.

   ii) It reports its opinion and recommendations on these matters together with the results of tests and trials it carries out and also the views and comments of the responsible technical establishments and officials who have been associated with investigations. These reports are recorded in its printed Proceedings as are also the statements of policy and decisions on these matters communicated to the Board by the Armament Directorates.

        Every member of the Board has the right to have recorded a minute of dissent from the opinion of the remainder of the Board.

   iii) For the above purposes the Board possesses the following powers:-

        a)   To ask the Research, Design and Experimental Establishments for any advice, designs, investigations and trials.

        b)   To consult any other official bodies or persons associated with scientific and technical development, including advisory Committees.

        c)   To stage and order trials.

        d)   To order the material required for those trials.

   iv) The Board has the right, in common with all other Establishments concerned with armament development, to submit to the approving technical Armament Directorate its own proposals and projects for the development or employment of armaments.

   v)   All technical Establishments concerned with armament development, such as the A.R.D. and A.D.D., can consult the Board direct.

22. We recommend:-
    i)   That the Ordnance Board should advise the appropriate ordering authorities on matters of conflicting experimental priorities as they arise and report to the Experimental Progress Committee.
    ii) That the Naval, General and Air Staffs be entitled to be represented at the meetings of the Ordnance Board if the Admiralty, War Office and Air Ministry so desire.
    iii) That a technical assistant, of Senior Technical Officer grade, be appointed for each of the main Committees of the Board, to give technical secretarial services to the Committees and to assist in preparing reports.
    iv) That an expert in office organisation be called in to investigate and overhaul the system of conducting office business at the Board and to make recommendations for its improvement. We expect that considerable changes will be necessary in staff and organisation of the secretariat, and a possible increase in printing facilities available to the Board.
    v)   That the mechanism of 'advanced action' should be resorted to whenever possible, and in order that this should not impede progress by further overlapping the printing machinery, facilities for typescript duplication of papers should be provided.

VI. Dispersal

23. Whatever justification existed for the original dispersal of A.R.D., A.D.D. and O.B. we consider the effects have been so serious that we recommend:-
    i)   That an heroic effort is imperative to reassemble them to a very considerable extent on one site within easy access of the centre of London, preferably to the West or North.
         The establishments which must be reassembled are the O.B. Headquarters of the A.R.D. and those sections of A.D.D. now at Knockholt and Halstead Place.
         Certain sections of A.R.D. and A.D.D. are already well located close to each other and with the advantage of close proximity to factories making the product, with which they are concerned. These considerations at the present time counter balance any advantage which might be obtained from their reassembly near other Sections and the O.B.
         The Sections of A.R.D. and A.D.D. at the small arms ammunition factory at Swynnerton and the small arms Section of A.D.D. at Cheshunt are cases in point which need not be disturbed.

ii) That each section of A.R.D. should be examined to see whether reassembly with A.R.D. H.Q., is justified and possible. In many cases the laboratories which have been created could be moved only with difficulty and when opportunity offers.

iii) That when the sections of A.D.D. at Knockholt and Halstead Place, the O.B. and H.Q. of A.R.D. are reassembled, the emergency provision to cover possible future War contingencies which may again compel dispersal.

Signed

H.L. Guy FRS	Chairman
Sir George Preece	Admiralty
Oliver Bevir	Admiralty
G.A.H. Pidcock	Air Ministry and MAP
P. Huskinson	Air Ministry and MAP
Edward Appleton FRS	Scientific Advisory Committee
H. Gough FRS	Ministry of Supply
E.M.C. Clarke	Ministry of Supply
A.F. Dobbie-Bateman (Dissenting however from Section V)	Ministry of Supply
L. Daryl Hickes	War Office

J.M.L. Ivimy   Secretary
12th August 1942

Appendix IX[1]

History of Anglo-American Relations in Atomic Energy up to the Quebec Agreement

---

1. The M.A.U.D. Committee first met, as a Sub-Committee of the Committee for Scientific Survey of Air Warfare (C.S.S.A.W.), on April 10th 1940 under the Chairmanship of Sir George Thomson, but it was not given these initials till the third meeting which was held on June 19th 1940. This Committee was originally responsible to the Air Ministry but was very soon transferred to the Ministry of Aircraft Production and, until the summer of 1941, it was responsible for the co-ordination and direction of all atomic energy research in the U.K.

2. At the fifth meeting of the M.A.U.D. Committee (August 7th 1940) it was decided that the forthcoming scientific mission to the U.S.A. should inform responsible American scientists of all that was being done in the atomic energy field in the U.K. and try to find out what was going on there.

3. The last two meetings of the M.A.U.D. Technical Sub-Committee were held on April 9th and July 2nd 1941. Dr Bainbridge of the American National Defence Research Committee (N.D.R.C.) was present at the first of these when a full review was made, for his benefit, of the current state of the work, and Professor Lauritsen, also of the N.D.R.C., was present at the second. The U.S. authorities were in this way, as well as through British scientific representatives in Washington who had copies of M.A.U.D. Committee minutes, fully informed of British views and progress and the Committee, in turn, knew of the corresponding American work.

4. The M.A.U.D. Committee presented its final report in July 1941 to the Director of Scientific Research in M.A.P. Earlier in the year he had

drawn attention to the rising cost of the work and asked for a considered recommendation and estimate of future requirements. It was understood that he proposed to forward the report to the Scientific Advisory Committee of the War Cabinet, under the Chairmanship of Lord Hankey, for consideration of the future of the work. The meeting of the M.A.U.D. Committee which Professor Lauritsen attended revised the final draft of this report.

5. Lord Cherwell also attended meetings of the M.A.U.D. Committee in order to keep the Prime Minister directly informed and reported the conclusions to him. As a result the Prime Minister sent the following minute to the Chiefs of Staff Committee on August 30th 1941, as quoted in the official 'Statements Relating to the Atomic Bomb':-

    General Ismay for Chiefs of Staff Committee.

    Although personally I am quite content with the existing explosives, I feel we must not stand in the path of improvement, and I therefore think that action should be taken in the sense proposed by Lord Cherwell, and that the Cabinet Minister responsible should be Sir John Anderson. I shall be glad to know what the Chiefs of Staff Committee think.

    The recommendations of the M.A.U.D. Committee were subsequently approved by the Scientific Advisory Committee of the War Cabinet.

6. Discussion took place during September between Sir John Anderson, Lord Brabazon (then Minister of Aircraft Production) and Lord McGowen on the transfer of the work and its future organisation. The M.A.U.D. Committee had recommended the granting of contracts by M.A.F. to I.C.I. in connection with the work, and discussions were in progress on the possibility that the Company should take over certain aspect for further large-scale development.

7. On October 15th 1941 Sir John Anderson saw Lord McGowen, Sir Wallace Akers and Mr. Perrin and, on the following day, Sir Wallace Akers and Sir Edward Appleton. Following these two meetings it was agreed that the Government (through D.S.I.R.) should take over all aspects of atomic energy work under the direct responsibility of the Lord President of Council. Sir Wallace Akers was to be released, part-time, from I.C.I. and Mr. Perrin whole-time to work for D.S.I.R.

    The Lord President of the Council was to be advised by a Consultative Council and Sir Wallace Akers, who had direct access to Sir John Anderson on all questions of policy, by a Technical Committee consisting of the leading scientists in the field.

On October 21st the new organisation was given the cover name of Directorate of Tube Alleys at the suggestion of Sir John Anderson. This seemed plausible and not too mysterious and tubes of one kind or another were likely to be involved in high priority jobs throughout the War.

8. On October 11th 1941 President Roosevelt wrote to the Prime Minister suggesting that they should soon correspond or converse 'in order that any extended efforts may be co-ordinated or even jointly conducted'. In December 1941 the Prime Minister replied: 'I need not assure you of our readiness to collaborate with the U.S. administration in this matter'.

9. Meanwhile Professors Pegram and Urey, who were Vice-Chairman and a member respectively of the American Committee of the N.D.R.C. which was then responsible for their atomic energy work, visited this country from October 27th till November 14th, and were taken to all the centres where work was in progress. They were given all details and there was full discussion on the organisations, programs and present state of the work in the two countries. Professors Pegram and Urey were present for part of the first meeting of the T.A. Technical Committee and undertook to discuss with their own authorities the possibility of transferring part of the U.K. program (under the direction of Dr H. Halban) to the U.S.A.

10. The Japanese attack on Pearl Harbour took place on December 7th 1941 but before this the Americans had strengthened their effort on the atomic energy problem and the 'uranium section' was taken out of the N.D.R.C. Development (O.S.R.D.) of which Dr Vannevar Bush was Director.

11. The policy of full exchange of information between the British and American teams was maintained during the first half of 1942. Sir Wallace Akers visited the U.S.A. between the end of January and the beginning of April and was accompanied for part of this time by Professor Peierls, Professor Simon and Dr Halban. Sir James Chadwick was, unfortunately, not able to join this team as had been proposed. During this time correspondence between Dr Bush and Sir John Anderson assumed on both sides complete collaboration at all stages of the project.

In June 1942 Mr. Perrin visited Canada and the U.S.A. The main object was to inform the Canadian Government of the Tube Alloy project and to discuss the possibility of their taking over the Eldorado Company which owned the uranium mines at Great Bear Lake. The Americans were informed of, and approved, this arrangement. During the visit there were further discussions with Dr Bush and Dr Conant on the possible integration of the British and American programs and with the Canadians on the possible transfer of Dr Halban and his team from Cambridge and the setting up of a joint project in Canada.

12. In June 1942 the Prime Minister visited Washington and had informal discussions on the atomic energy project with President Roosevelt at Hyde Park. He reported that the whole basis of the conversation was that there was to be complete co-operation and sharing of results. (The fall of Tobruk was announced just after the Prime Minister arrived in Washington.).

13. When Mr. Perrin had returned to London and reported to Sir John Anderson the latter had an extensive correspondence with Dr Bush on the possible division of scientific and industrial resources between the two countries so as to provide the maximum help in the realisation of the project and to obtain an effective Government control which could be continued after the War.

   Negotiations were also carried on, with Dr Bush's knowledge, through Mr. M. MacDonald on the transfer of part of the British program to Montreal. This was finally agreed at a meeting in London on October 12th 1942 at which Mr. C.D. Howe and Mr. MacDonald were present.

14. At the end of that month Sir Wallace Akers visited Canada and the U.S.A. again in order to make detailed arrangements for setting up the joint Canadian project and to carry further the correspondence between Sir John Anderson and Dr Bush in the matter of an integrated program with the U.S.A.

   On his arrival it became clear that the U.S. Army had taken control of the atomic energy project and that the O.S.R.D. organisation was only concerned with some of the research and development aspects. One consequence of this was a tightening of the secrecy regulations and a strict application of the principle of 'compartmentalisation' by which only those directly concerned knew about any particular part of the project and there was very little general knowledge of the whole.

15. Sir Wallace Akers was asked to wait till 'top level' directives, which were going to be issued, had made clear the extent to which information could be exchanged. Eventually he received a message from Dr Conant on January 7th 1943 that the directive had been received and went to Washington where he was handed a memorandum on January 13th.

   This laid down the conditions for interchange of information with the British and Canadians which were governed by the basic principle 'that interchange on design and construction of new weapons and equipment is to be carried out only to the extent that the recipient of the information is in a position to take advantage of this information in this War.'

16. It was recognised at once that the logical application of this principle would seriously affect the policy of full co-operation with the U.S.A. which had, until this time, underlain the atomic energy project and would make the position of the team which had just started work in Montreal extremely difficult.

    The information was cabled to the Prime Minister who had gone to the meeting with President Roosevelt at Casablanca (January 20th 1943). The latter expressed his general approval of collaboration in the atomic energy field and promised to investigate the position on his return to the U.S.A.

17. On February 26th 1943 the Prime Minister received a telegram from Mr. Hopkins saying that the American authorities did not realise that any agreement had been broken and asking for details of the British case. The reply, which was sent on February 27th and which asked for a definite statement of policy, included a memorandum giving a history of U.S.-U.K. relations on the atomic energy project up to the time of the Conant Memorandum.

    In spite of several reminders and an interim telegram from Mr. Hopkins on March 19th that 'I am working on Tube Alloys and will let you know as soon as I know something definite', no definite reply was ever received.

18. In April considered reports were submitted to Sir John Anderson by the Directorate of Tube Alloys on the results that would follow acceptance of the Conant Memorandum and on the effort that would be involved in continuing a full-scale program in the U.K. and in Canada if there was no collaboration with the U.S.A. The latter was prepared in answer to a minute from the Prime Minister to the Lord President of the Council dated April 15th and was sent to him on April 19th after discussion at a meeting of the Consultative Council.

19. In May 1943 the Prime Minister again visited Washington and was accompanied by Lord Cherwell. The latter had a preliminary discussion with Dr Bush and Mr. Hopkins in which he represented the U.K. view that we must, as a nation, continue atomic energy work. It would be much more efficient to work jointly but the U.K. Government must be an equal partner and could not accept a position as one 'cell' in an American organisation. Lord Cherwell left Washington before the Prime Minister

discussed the issue with the President on May 26th. This conversation was recorded in the following terms:-

> The President agreed that the exchange of information on Tube Alloys should be resumed and that the enterprise should be considered a joint one, to which both countries would contribute their best endeavours. I understood that his ruling would be based upon the fact that this weapon may well be developed in time for the present War and that in thus falls within the general agreement covering an interchange of research and invention secrets.

20. In the expectation of early action Sir Wallace Akers again went to Canada in June 1943 so as to be ready to resume discussions with the Americans. On June 10th the Prime Minister telegraphed to Mr. Hopkins reminding him of his conversation with the President and asking him to telegraph as soon as his instructions had been issued. On July 9th he telegraphed to the President as follows:-

> Since Harry's telegram of June 17th I have been anxiously awaiting further news about Tube Alloys. My experts are standing by and I find it increasingly difficult to explain the delay. If difficulties have arisen I beg you to let me know at once what they are in case we may be able to help in sorting them.

21. In July 1943 Mr. Stimson and Dr Bush came to London and had discussions on the Tube Alloys problem with the Prime Minister, Sir John Anderson and Lord Cherwell. In these meetings the Heads of a possible agreement between the two Governments were discussed and on July 27th the Prime Minister received a telegram from the President saying that everything was satisfactorily arranged and suggesting a visit from the 'top man' to Washington to complete details.

Accordingly Sir John Anderson went to Washington from August 1st to the 13th where he was joined by Sir Wallace Akers. In meetings with Dr Bush and Dr Conant the precise form of an agreement governing exchange of information between the U.K. and the U.S.A. was settled and this was signed by the Prime Minister and the President during the Quebec Conference which took place at the time. The Prime Minister of Canada also accepted the agreement.

Appendix X[1]

Briefing Papers for the Dambusters Raid

Most Secret

No 5 Group Operation order No B.976
Appendix A routes and timings
Appendix B signals procedure for target diversions etc.
Appendix C light and moon table

Information

General

1. The inhabitants and industry of the Ruhr rely to a very large extent on the enormously costly water barrage dams in the Ruhr District. Destruction of TARGET X alone would bring about a serious shortage of water for drinking purposes and industrial supplies. This shortage might not be immediately apparent but would certainly take effect in the course of a few months. The additional destruction of one or more of the five major dams in the Ruhr Area would greatly increase the effect and hasten the resulting shortage. TARGET Z is next in importance.

2. A substantial amount of damage would be done, and considerable local flooding would be caused immediately consequent on the breach of TARGET X. In fact it might well cause havoc in the Ruhr valley. There would be a large loss of electrical capacity in the Ruhr partly caused by destruction of hydro-electric plants, but also due to loss of cooling water for the large thermal plants.

3. In the Weser District the destruction of the TARGET Y would seriously hamper transport in the Mittelland Canal and in the Weser, and would probably lead to an almost complete cessation of the great volume of traffic now using these waterways.

4. The reservoirs usually reach their maximum capacity in May or June, after which the level slowly falls.

Enemy Defences

5   (a) TARGET X
There are three objects on the crest of this dam which may each be a light A.A. gun. A light 3-gun A.A. position is situated below and to the north of the dam with a possible searchlight position nearby. A double line boom with timber spreaders is floating on the main reservoir at 100 to 300 feet from the dam. No other A.A. position or defence installation is known.

(b) TARGETS Y and Z
Information about the defences of these two dams will be given when P.R.U. sorties have covered these areas (information has now been issued).

(c) The last resort targets are unlikely to be defended.

Intention

To breach the following dams in order of priority as listed:
a) TARGET X   (GO 939)
b) TARGET Y   (GO 934)
c) TARGET Z   (GO 960)
d) Last resort targets:
    i)   TARGET D   (GO 938)
    ii)   TARGET E   (GO 935)
    iii) TARGET F   (GO 933)

Execution

Code Name

7. This operation will be known by the code name which will be issued separately.

Date of attack

8. The operation is to take place on the first suitable date after 15th May 1943.

Effort

9. Twenty special Lancasters from 617 Squadron.

Outline Plan

10. The twenty special Lancasters of 617 Squadron are to fly from base to
    target area and return in moonlight at low level by the routes given in
    Appendix A. The Squadron is to be divided into three main waves, viz:
    a) 1st Wave. Is to consist of three sections, spaced at ten minute inter-
        vals, each section consisting of three aircraft. They are to take the
        Southern route to the target area and attack TARGET X. The attack
        is to be continued until the dam has been clearly breached. It is esti-
        mated that this might require three effective attacks. When this has
        been achieved the leader is to divert the remainder of this wave to
        TARGET Y, where similar tactics are to be followed. Should both X
        and Y be breached any remaining aircraft of this wave are to attack
        Z.
    b) 2nd Wave. Is to consist of five aircraft manned by the specially
        trained crews who are to take the Northern route to the target, but
        are to cross the enemy coast at the same time as the leading section
        of the first wave. This 2nd wave are to attack TARGET Z.
    c) 3rd Wave. Is to consist of the remaining aircraft and is to form an
        airborne reserve under the control of Group H.Q. They are to take
        the Southern route to the target but their time of take off is to be
        such that they may be recalled before crossing the enemy coast if the
        1st and 2nd waves have breached all the targets.
        Recall will probably not be possible unless the first section of the 1st
    Wave are at position 51°51'N, 03°00'E by Civil Twilight (Evening) + 30
    minutes and the 3rd Wave must be at this position 2 hours 30 minutes
    later. Orders will be passed to aircraft on the Special Group frequency
    if possible before they reach the enemy coast instructing them which
    target they are to attack. Failing receipt of this message the aircraft are to
    proceed to X, Y and finally last resort targets in that order, attacking any
    which are not breached. Officer Commanding, RAF Station, Scampton,
    is to arrange for individual aircraft to be detailed to specific last resort
    targets.

Detailed plan

11. The 1st Wave is to take off in three sections each of three aircraft and fly
    to the target at low level by the route given in Appendix A. Sections are
    to be spaced at intervals of ten minutes and are to fly in open formation.
    Height is not to exceed 1,500 feet over England. On leaving the English
    Coast aircraft are to descend to low level and set their altimeters to 60
    feet using the Spotlight Altimeters for calibration. The QFF at various

stages of the route is to be carefully noted. Aircraft are to remain at low level for the Flight to the target and on the return journey at least until crossing a point 03°00'E.

12. An accurate landfall on the enemy coast is important but on no account should the aircraft turn back if their landfall is not quite accurate. The routes selected should be free of all major opposition from flak but a good map reading and crew co-operation is essential to keep aircraft on track. The enemy coast is to be crossed as low as possible going in and coming out even if it is necessary to climb a little later for map reading.

13. On arriving at a point 10 miles from the target the leader of each section is to climb to about 1,000 feet. On seeing this the other aircraft are to listen out on V.H.F. Each aircraft is to call the leader of the Wave on V.H.F. on arriving at the target. Spinning of the special store is to be started ten minutes before each aircraft attacks. The leader is to attack first and is then to control the attacks on TARGETS X and Y by all the other aircraft of the 1st Wave using the signal procedure given in appendix B.

14. Number 2 of the leading section of the 1st Wave is to act as deputy leader for the whole of the 1st Wave during the attack on TARGET X. Should the leader fall out No 2 of the leading section is to take over the leadership, and No 3 deputy leadership, for the attack on TARGET X. For the attack on TARGET Y Number 4 is to take over deputy leadership, in which event No 7 is to be the deputy leader. All other aircraft are to return to base after completing their attack. The first three aircraft are to return by Route 1, the second three by Route 2 and the last three of this Wave by Route 3.

15. The direction of attack of TARGET X is to be at right angles to the length of the target. The general direction of attack is, therefore, to be S.E. to N.W. Aircraft are not to be diverted to TARGET Y until TARGET X has been breached. If TARGET X is breached, up to two additional aircraft may be used at the discretion of the leader, to widen the breach in TARGET X providing at least three aircraft are diverted to attack TARGET Y.

16. Destruction of the Dam may take some time to become apparent and careful reconnaissance may be necessary to distinguish between breaching of the dam and the spilling over the top, which may follow each explosion.

17. When TARGET X is seen to be breached beyond all possible doubt the leader is to divert the remainder of the first Wave to TARRGET Y by W/T and V.H.F. where similar tactics are to be used for the attack of this target. The general direction of attack of TARGET Y is to be from N.W. to S.E. If TARGET Y is seen to be breached beyond all possible doubt all remaining aircraft of the 1st Wave are to be diverted by the leader to attack TARGET Z independently using the same tactics as the 2nd Wave.

18. For the attacks of both TARGETS X and Y the special range finder is to be used, the height of attack is to be 60 feet and the ground speed 220 mph.

19. The 2nd Wave is to take off and fly to TARGET Z at low level by the Northern Route given in Appendix A. Aircraft are to cross the enemy coast in close concentration, but not in formation, at the same time, although at a different point, as the leading section of the 1st Wave. Aircraft of this Wave will be controlled on the alternative V.H.F. channel. The special stores are not to be spun for the attack of TARGET Z. Aircraft are to attack this target from N.W. to S.E. parallel to the length of the dam and are to aim to hit the water just short of the centre point of the dam about 15 to 20 feet out from the edge of the water. Attacks are to be made from the lowest practicable height at a speed of 180 mph. I.A.S. Aircraft are to return to base independently. First two aircraft by Route 1; second two aircraft by Route 2 and the last by Route 3.

20. The 3rd Wave is to consist of the remaining aircraft and is to form an airborne reserve under the control of Group Headquarters. They are to fly to TARGET X in close concentration, but not in formation, at low level by the Southern route given in Appendix A. These aircraft are to be at position 51°52'N, 03°00'E 2 hours and 30 minutes after the leading section of the 1st Wave have crossed this point on their outward route to the target. Orders for the 3rd Wave will be passed to all aircraft on the special Group frequency, if possible before they reach the enemy coast, instructing them which target they are to attack. Failing receipt of this message aircraft are to proceed to X, Y and finally, last resort targets in that order attacking any which are not breached. The 3rd Wave are to use tactics of attack similar to those used by the 1st Wave when attacking TARGETS X and Y except that attacks on last resort targets are to be made independently. After attacking, aircraft are to return to base independently at low level by any of the three return routes given in Appendix

A. Aircraft attacking early should take Route 1, next aircraft Route 2 and the last Route 3.

Method of Attack

21. Aircraft are to use the method of attack already practiced. The pilot being responsible for line, the Navigator for height, the Air Bomber for range and the Flight Engineer for speed.

22. The interval between attacking aircraft is to be not less than three minutes on all targets.

23. On all targets except TARGET Z each aircraft is to fire a red verey cartridge immediately over the dam during the attack. Aircraft attacking TARGET Z are each to fire a red verey cartridge as they release their special store.

24. All aircraft are to fly left hand circuits in each target area keeping as low as possible when waiting their turn to attack.

Time of Attack

25. The time of attack of each target by each wave is not important to within a few minutes. The time of crossing the enemy coast is, however, all important. ZERO HOUR, which will be given in the executive order, is therefore, to be the time at which the first section of the 1st Wave are to be at position 51°52'N, 03°00'E on the outward route to the target. This will probably be Civil Twilight (Evening) + 30 minutes. At this time aircraft of the 2nd Wave should be about position 53°19'N, 04°00'E.

Routes

26. As in Appendix A.

Diversions

27. The whole essence of this operation is surprise, and to avoid bringing enemy defences to an unnecessary degree of alertness, diversionary attacks must be carefully timed. H.Q.B.C. will be asked to arrange the maximum possible diversionary attacks so that the first enemy R.D.F. or other warning of the diversionary attacks occurs 20 minutes after the leading section of the 1st Wave crosses the enemy coast. No diversionary

attacks should be dispatched which would cross the enemy coast for a period of one hour preceding the 3rd Wave. 15 minutes after the 3rd Wave cross the enemy coast further diversionary attacks should be made at maximum strength and should continue, if possible until the 3rd Wave are clear of enemy territory on the return journey. Diversionary attacks below 2,000 feet should not be made in the area bounded by the points (51°00'N, 03°20'E), (51°20'N, 06°30'E), (51°00'N, 09°00'E) and (53°20'N, 06°00'E). H.Q.B.C. will also be asked to arrange suitable weather reconnaissance to report in particular on the visibility in the target area at least in sufficient time to recall the Lancasters before they cross the enemy coast if the weather is unsuitable.

Armament

28.
   a) Bomb Load – Each Lancaster is to carry one special modified store (UPKEEP).
   b) Ammunition – all guns to be loaded with 100% night tracer (G VI).

Fuel

29. The Lancasters may take off at a maximum all up weight of 63,000 lbs at +14boost. As the modified store now weighs about 9,000 lbs, 1750 gallons of petrol can be carried.

Navigation

30. H.Q.B.C. are requested to arrange for the Eastern Chain, Stud 5 to be switched on as Z-20minutes and to remain on for the whole operation. This should assist in making an accurate landfall on the enemy coast at the correct time.

31. The route is to be carefully studied before the flight and the outstanding features, obstructions and pinpoints noted, particularly water pinpoints. E.T.A.'s at each are to be carefully calculated and if any pinpoint is not found on E.T.A. a search is to be made before proceeding to the next pinpoint. Aircraft may climb to 500 feet shortly before reaching each pinpoint if necessary to help map reading.

32. The maximum use is to be made of the Air Position Indicators.

Synchronisation of watches

33. All watches are to be synchronised with BBC time before take-off on the day of the operation.

Secrecy

34. Secrecy is VITAL. Knowledge of this operation is to be confined to the Station Commander, O.C. 617 Squadron and his two Flight Commanders until receipt of the Executive signal. After crews are briefed they are to be impressed with the need for the utmost secrecy because of the possibility that the operation may be postponed should weather reconnaissance prove the weather to be unsuitable.

Reports

35. Each aircraft as soon as possible after it has attacked is to report by W/T on the normal Group operational frequency in accordance with Appendix B.

Special Devices

36. MANDREL and TINSEL are not fitted.

37. IFF is NOT to be used on the outward journey by normal procedure is to be followed on the homeward flight. Any aircraft returning early is NOT to use IFF except after Z+30 minutes for the 1st and 2nd Waves and after Z+3 hours for the 3rd Wave.

Nickels

38. Nickels are not to be dropped.

Intercommunication

Wireless Silence

39. Strict W/T and R/T silence is to be maintained until after Z+30 minutes for the 1st and 2nd Waves and after Z+3 hours for the 3rd Wave. Any aircraft returning early is NOT to break W/T or R/T silence and is NOT to identify on MF/DF except after Z+30 minutes for the 1st and 2nd Waves and after Z+3 hours for the 3rd Wave. Aircraft returning before that time are to cross the English coast at 1,500 feet at the point of exit and proceed direct to base or the nearest suitable airfield. Otherwise

normal operation signals procedure is to be used except as modified by Appendix B.

MF/DF Section

40. Section D is to be used if required in accordance with paragraph 39.

Executive Order

41. The executive order for the operation will be given by EXECUTIVE followed by the code word allotted, the date on which the operation is to take place and the time of Zero Hour in British Double Summer Time.

42. ACKNOWLEDGE BY TELEPRINTER

Signed Satterly
Senior Air Staff Officer
No 5 Group
Royal Air Force
16th May 1943

Appendix XI[1]

Sub-Committees and Panels of the Aeronautical Research Committee

---

Membership and terms of reference of The Aeronautical Research Committee its Sub-Committees and Panels
1st May 1940

Aeronautical Research Committee
Terms of reference

1) To advise the Secretary of State on scientific problems relating to aeronautics.
2) To make from time to time recommendations to the Air Council as to any researches which the Committee consider it desirable to initiate, and to any matters referred to them by the Council.
3) To supervise the aeronautical researches at the National Physical Laboratory initiated by them and if requested to do so by the Air Council any other researches connected with Aeronautics.
4) To make an annual report to the Air Council of the research work which the Committee consider should be undertaken at the National Physical Laboratory, or elsewhere, together with an estimate of expenditure at the National Physical Laboratory.
5) To investigate the causes of such accidents as may be referred to them by the Air Council and to make recommendations as to the prevention of accidents in the future.
6) To promote education in aeronautics by co-operating with the Governors of the Imperial College and in any other way within their power.
7) To assist with advice any research carried out by or on behalf of the Aeronautical Industry and to make available any information of value to the Industry so far as is compatible with public interest.
8) To make an annual report to the Secretary of State for Air.

Aerodynamics Sub-Committee

Terms of reference
1) To suggest and supervise general researches on aerodynamics at the National Physical Laboratory.
2) To advise on aerodynamic researches carried out in the Research Establishments of the Air Ministry.
3) To report to the Aerodynamic Research Committee.

Independent Members:
Sir Leonard Bairstow FRS (Chairman), Lord Blackett FRS, Sir Bennett Jones FRS, Professor W.J. Duncan FRS, Professor G.T.R. Hill, Sir Geoffrey Taylor FRS, Sir George Thomson FRS, Sir Henry Tizard FRS

Official Members:
Dr G.P Douglas (R.A.E.), Mr A. Fage FRS (N.P.L.), Sir William Farren FRS (A.M.), Sir David Pye FRS (A.M.), Mr E.F. Relf FRS (N.P.L.), Dr H.C.H. Townend (Admiralty), Dr H. Roxbee Cox (A.M.), Mr H.F. Vessey (A.M.) and Mr J.L. Nayler (N.P.L.)

Alloys Sub-Committee

Terms of Reference

1) To suggest and supervise general researches on alloys for aeronautical purposes at the National Physical Laboratory.
2) To advise on metallurgical researches carried out at the Research Establishments of the Air Ministry.
3) To consider any matters relating to the improvement of alloys for aeronautical purposes referred to them by the Main Committee.
4) To report to the Aeronautical Research Committee.

Independent Members:
Dr L. Aitchison, Sir William L. Bragg FRS, Dr C.H. Desch FRS, Dr H. Moore, Dr R. Seligman, Sir Geoffrey Taylor FRS, Sir Henry Tizard FRS

Official Members:
Dr C. Coldron-Smith (Admiralty), Dr H.J. Gough FRS (War Office), Dr R.H. Greaves (War Office), Mr A.P. Patterson (Admiralty), Sir David Pye FRS (A.M.), Dr S.L. Smith (N.P.L.), Dr H. Sutton (R.A.E.), Sir Charles Sykes FRS (N.P.L.), Mr A.H. Waterfield (A.M.), Dr M.L. Becker (N.P.L.) Secretary

Elasticity and Fatigue Sub-Committee

Terms of Reference

1) To suggest and supervise researches on elasticity and fatigue undertaken at the National Physical Laboratory.
2) To advise on elasticity and fatigue researches undertaken at Air Ministry Establishments.
3) To report to the Aeronautical Research Committee.

Independent Members:
Sir Geoffrey Taylor FRS (Chairman), Dr R.W. Bailey, Sir Ralph Fowler FRS, Professor A. Robertson, Sir Richard Southwell FRS, Sir Henry Tizard FRS

Official Members:
Dr W.D. Douglas (R.A.E.), Dr H.J. Gough FRS (War Office), Sir Ben Lockspeiser FRS (A.M.), Sir David Pye FRS (A.M.), Dr S.L. Smith (N.P.L.), Dr H. Sutton (R.A.E.), Sir Charles Sykes FRS (N.P.L.), Mr H.L. Cox (N.P.L.) Secretary

Engine Sub-Committee

Terms of Reference

1) To suggest and supervise researches on aircraft engine problems at the National Physical Laboratory and at various universities.
2) To advise on engine researches carries out at the Royal Aircraft Establishment and the Air Ministry Laboratory.
3) To consider problems on aircraft engines referred to them by the Main Committee.
4) To report to the Aeronautical Research Committee.

Independent Members:
Sir Henry Tizard FRS, Sir Leonard Bairstow FRS, Sir Alfred Egerton FRS, Major F.M. Green, Sir Harry Ricardo FRS

Official Members:
Major G.P. Bulman (A.M.), Mr H. Constant FRS (R.A.E.), Captain the Hon. D.C. Maxwell (Admiralty), Sir David Pye FRS (A.M.), Major A.A. Ross (A.M.), Mr W.L. Tweedie (A.M.), Mr A.F.C. Brown (N.P.L.) Secretary

Fleet Air Arm Research Sub-Committee

Term of Reference

To study the basic problems of the Fleet Air Arm and to advise on the research needed to solve these problems.

Independent Members:
Sir Leonard Bairstow FRS, Sir George Thomson FRS, Sir Henry Tizard FRS

Official Members:
Mr J.H. Chapman (Admiralty), Captain R.M. Ellis (Admiralty), Commander A.C.G. Ermen (Admiralty), Sir William Farren FRS (A.M.), Mr J.A.C. Manson (A.M.), Sir David Pye FRS (A.M.), Mr E.F. Relf FRS (N.P.L.), Mr C.S. Wright (Admiralty), Mr J.L. Naylor (N.P.L.) Secretary

Metrology Sub-Committee

Terms of reference

1) To consider those meteorological problems which affect the flying of aircraft.
2) To report to the Aeronautical Research Committee.

Independent Members:
Dr G.M.B. Dobson FRS (Chairman), Sir David Brunt FRS, Sir Bennett Jones FRS, Sir Geoffrey Taylor FRS, Sir Henry Tizard FRS

Official Members:
Dr N.K. Johnson (A.M.), Sir Ben Lockspeiser FRS (A.M.), Sir David Pye FRS (A.M.), Dr J.E. Ramsbottom (R.A.E.). Mr E. Ower (R.A.E.) Secretary

Oscillation Sub-Committee

Terms of Reference

1) To discuss problems on the flutter of aeroplanes and allied matters, including oscillations and vibrations and such other matters as may be referred to them by the A.R.C. from time to time.
2) To report to the Aeronautical Research Committee.

Independent Members:
Professor W.J. Duncan FRS (Chairman), Sir Leonard Bairstow FRS, Sir Richard Southwell FRS, Sir Geoffrey Taylor, Sir Henry Tizard FRS (ex officio)

Official Members:
Dr R.A. Frazer FRS (N.P.L.), Mr H. Grinsted (A.M.), Sir Alfred Pugsley FRS (R.A.E.), Sir David Pye FRS (A.M.), Dr D Williams (A.M.), Mr A.R. Collar (N.P.L.) Secretary

Plastics Sub-Committee

Terms of reference

1) To consider the properties of composite and reinforced materials with special reference to stressed components suitable for the construction of aircraft.
2) To report to the Aeronautical Research Committee.

Independent Members:
Sir Ralph Fowler FRS (Chairman), Sir Richard Southwell FRS, Sir Geoffrey Taylor FRS, Sir Henry Tizard FRS

Official Members:
Dr H. Roxbee (R.A.E.), Sir Charles Darwin FRS (N.P.L.), Dr W.D. Douglas (R.A.E.), Mr J. Latham (F.P.R.L.), Mr N.J.L. Megson (War Office), Dr K.W. Pepper (C.R.L.), Sir David Pye FRS (A.M.), Dr S.L. Smith (N.P.L.), Mr A.H. Waterfield (A.M.), Mr H.L. Cox (N.P.L.) Secretary

Seaplane Sub-Committee

Terms of Reference

1) To suggest and supervise researches on seaplane investigations under-taken at the National Physical Laboratory.
2) To advise on seaplane researches carried out at the Royal Aircraft Establishment and at the Marine Aircraft Experimental Establishment.
3) To report to the Aeronautical Research Committee.

Independent Members:
Lord Blackett FRS (Chairman), Sir Leonard Bairstow FRS, Mr A.M. Binnie FRS, Sir Henry Tizard FRS, Mr H.E. Wimperis

Official Members:
Dr G.S. Baker (N.P.L.), Sir William Farren FRS (A.M.), Mr H.M. Garner (M.A.E.E.), Mr A.J.T. Gibbons (Admiralty), Dr J.P. Gott (R.A.E.), Major R.E. Penny (A.M.), Sir David Pye FRS (A.M.), Mr C.S. Wright (Admiralty), Mr J.L. Nayler (N.P.L.) Secretary

Stability and Control Sub-Committee

Terms of Reference

1)   To suggest and supervise researches at the National Physical Laboratory on the stability and control of aeroplanes, with particular reference to control at low speeds.
2)   To advise on researches carried out at Air Ministry Establishments on the stability and control of aeroplanes, with particular reference to control at low speeds.
3)   To report to the Aeronautical Research Committee.

Independent Members:

Sir George Thomson FRS (Chairman), Sir Leonard Bairstow FRS, Dr H Roxbee Cox, Professor W.J. Duncan FRS, Professor G.T.R. Hill, Sir Bennett Jones FRS

Official Members:

Mr L.W. Bryant (N.P.L.), Sir William Farren FRS (A.M.), Mr S.B. Gates FRS (R.A.E.), Mr H.B. Irving (R.A.E.), Mr E.T. Jones (A.&A.E.E.), Squadron Leader J.F.X. McKenna (A.M.), Sir David Pye FRS (A.M.), Mr E.F. Relf FRS (N.P.L.), Mr N.E. Rowe (A.M.)

Structure Sub-Committee

Terms of Reference

1)   To consider problems relating to the structural design of aircraft; to advise on the initiation of research both theoretical and experimental; and to consider and advise on structural research in progress at the National Physical Laboratory and at the Royal Aircraft Establishment.
2)   To co-ordinate, in consultation with the Elasticity and Fatigue Sub-Committee, researches on aircraft materials as these affect structural design.

3)   To report to the Aeronautical Research Committee.

Independent Members:
Sir Ralph Fowler FRS (Chairman), Sir Leonard Bairstow FRS, Dr H. Roxbee Cox, Professor W.J. Duncan FRS, Professor G.T.R. Hill, Professor A.J. Sutton Pippard FRS, Sir Richard Southwell FRS, Sir Henry Tizard FRS

Official Members:
Mr H.L Cox (N.P.L.), Dr W.D. Douglas (R.A.E.), Mr H. Grinsted (A.M.), Sir Ben Lockspeiser FRS (A.M.), Sir Alfred Pugsley FRS (R.A.E.), Sir David Pye FRS (A.M.), Dr D. Williams (A.M.), Mr H.L. Cox (N.P.L.) Secretary

Free Flight Panel

Terms of Reference

1)   To discuss methods of making simpler stability experiments on aeroplane models.
2)   To report to the Aeronautical Research Committee.

Independent Members:
Sir Leonard Bairstow FRS (Chairman), Professor G.T.R. Hill, Sir Geoffrey Taylor FRS, Sir George Thomson FRS, Sir Henry Tizard FRS (ex officio)

Official Members:
Sir William Farren FRS (A.M.), Mr E.F. Relf FRS, (N.P.L.), Mr E. Ower (N.P.L.)

Navigation Panel

Terms of Reference

1)   To examine present methods of navigation by sextant and to consider any suggestions for improvements, particularly to methods for reducing observations.
2)   To report to the Aeronautical Research Committee.

Independent Members:

Sir Geoffrey Taylor FRS (Chairman), Instructor-Captain T.Y. Baker, Sir Bennett Jones FRS, Sir Henry Tizard FRS (ex officio)

Official Members:
Sir Ben Lockspeiser FRS (A.M.), Wing Commander F.M.V. May (A.M.),
Sir David Pye FRS (A.M.), Mr W.J. Richards (R.A.E.), Mr C.S. Wright
(Admiralty), Mr J.L. Nayler (N.P.L.) Secretary

R.A.E. High Speed Wind Tunnel Panel

Terms of Reference

1) To advise on the design and operation of the proposed high speed wind
   tunnel to be erected at the Royal Aircraft Establishment.
2) To report to the Aeronautical Research Committee from time to time on
   advice tendered by the Panel direct to the Air Ministry.

Independent Members:
Sir Bennett Jones FRS (Chairman), Sir Leonard Bairstow FRS, Sir Geoffrey
Taylor FRS

Official Members:
Dr G.P. Douglas (R.A.E.), Sir William Farren FRS (A.M.), Mr A.H. Hall
(R.A.E.), Mr W.G.A. Perring (R.A.E.) Secretary, Mr E.F. Relf FRS (N.P.L.)

Airscrew Panel

Terms of Reference

1) To suggest and supervise airscrew researches undertaken at the National
   Physical Laboratory.
2) To advise on airscrew researches undertaken by Air Ministry
   Establishments.
3) To report to the Aerodynamics Sub-Committee.

Independent Members:
Sir Leonard Bairstow FRS (Chairman), Dr S. Goldstein FRS

Official Members:
Mr C.N.H. Lock (N.P.L.), Mr W.G.A. Perring (R.A.E.), Mr H.F. Vessey
(A.M.) Mr A.R. Collar FRS (N.P.L.) Secretary

Fluid Motion Panel

Terms of Reference

1) To consider the development of the mathematical theory of real fluids in relation to the problems of aeronautics, and to advise on the initiation of related research, both theoretical and experimental.
2) To report to the Aerodynamics Sub-Committee.

Independent Members:
Sir Leonard Bairstow FRS (Chairman), Professor W.G. Bickley, Dr S. Goldstein FRS, Sir Bennett Jones FRS, Sir Richard Southwell FRS, Sir Geoffrey Taylor FRS

Official Members:
Mr A. Fage FRS (N.P.L.), Mr E.F. Relf FRS (N.P.L.), Mr L.F.G. Simmons (N.P.L.), Professor H.B. Squire FRS (A.M.), Mr A.D. Young FRS (R.A.E.), Mr E. Ower (N.P.L.) Secretary

Kite Balloon Panel

Terms of Reference

To consider such problems relating to kite balloons and kites as are referred to them.

Independent Members:
Sir Leonard Bairstow FRS (Chairman), Professor G.T.R. Hill, Sir Richard Southwell FRS, Sir Geoffrey Taylor FRS

Official Members:
Mr H. Bateman (A.M.), Mr L.W. Bryant (N.P.L.), Dr H. Roxbee Cox (A.M.), Sir Ben Lockspeiser FRS (A.M.), Mr E.F. Relf FRS (N.P.L.), Mr E. Ower (N.P.L) Secretary

Lubrication Panel

Terms of Reference

1) To consider problems of lubrication of internal combustion engines.
2) To suggest and advise on lines of research with a view to:

a) The elimination of unfavourable characteristics from the oils them-
    selves, or

b) The elimination of the evil effects resulting from those characteristics.

3) To review the possibility of developing more reliable methods of
    specification.

4) To report to the Engine Sub-Committee.

Independent Members:
Sir Henry Tizard FRS (Chairman), Sir Alfred Egerton FRS, Professor A.H.
Gibson, Sir Harry Ricardo FRS

Official Members:
Dr J.E. Ramsbottom (R.A.E.), Dr S.L. Smith (N.P.L.), Mr W.L Tweedie
(A.M.), Mr A.F.C. Brown (N.P.L.) Secretary

Notes

Chapter 1

1   *Notes and Records*, The Royal Society, 1940–41, p. 7.
2   The Royal Society, *Officers* <http://royalsociety.org/about-us/governance/officers/>. (Accessed April 2013).
3   H.A. Brück, *The Story of Astronomy in Edinburgh* (Edinburgh: Edinburgh University Press, 1983), p.84.
4   Joseph Whitaker, *Whitaker's Almanack 1941* (Whitaker, 1941), p. 291.
5   *Biographical Memoirs of Fellows of the Royal Society* Vol 1 1955, pp. 19–20.
6   *Obituary Notices of Fellows of the Royal Society* Vol 8 1952–1953, p. 271.
7   Ibid., p. 301.
8   *Obituary Notices of Fellows of the Royal Society* Vol 7 1950–1951, p. 32.
9   *Biographical Memoirs of Fellows of the Royal Society* Vol 4 1958, p. 58.
10   *Biographical Memoirs of Fellows of the Royal Society* Vol 16 1970, p. 43.
11   Ibid., p. 61.
12   Joseph Whitaker, *Whitaker's Almanack 1941*, p. 317.
13   Ibid., p. 320.
14   Ibid., p. 354.
15   Ibid., p. 355.
16   Ibid., p. 358.
17   *Biographical Memoirs of Fellows of the Royal Society* Vol 21 1975, pp. 447–84.
18   *Biographical Memoirs of Fellows of the Royal Society* Vol 57 2011, pp. 330–1.
19   Council for Assisting Refugee Academics, *Academic Freedom* <http://www.academic-refugees.org/>. (Accessed April 2013).
20   Bodleian Library University of Oxford, *Catalogue of the Archive of the Society for the Protection of Science and Learning* (1933–87) <http://www.bodley.ox.ac.uk/dept/scwmss/wmss/online/modern/spsl/spsl.html>. (Accessed April 2013).
21   *Biographical Memoirs of Fellows of the Royal Society* Vol 50 2004, p. 235.
22   *Biographical Memoirs of Fellows of the Royal Society* Vol 51 2005, p. 152.
23   *Biographical Memoirs of Fellows of the Royal Society* Vol 45 1999, p. 522.
24   *Biographical Memoirs of Fellows of the Royal Society* Vol 54 2008, p. 178.
25   *Biographical Memoirs of Fellows of the Royal Society* Vol 48 2002, p. 380.

26   *Biographical Memoirs of Fellows of the Royal Society* Vol 18 1972, p. 414.
27   *Biographical Memoirs of Fellows of the Royal Society* Vol 17 1971, p. 448.
28   *Biographical Memoirs of Fellows of the Royal Society* Vol 13 1967, p. 84.
29   *Biographical Memoirs of Fellows of the Royal Society* Vol 52 2006, p. 142.
30   *Biographical Memoirs of Fellows of the Royal Society* Vol 54 2008, p. 322.
31   *Biographical Memoirs of Fellows of the Royal Society* Vol 16 1970, p. 411.
32   *Biographical Memoirs of Fellows of the Royal Society* Vol 53 2007, pp. 49–50.
33   *Biographical Memoirs of Fellows of the Royal Society* Vol 29 1983, pp. 227–44.
34   *Biographical Memoirs of Fellows of the Royal Society* Vol 51 2005, p. 459.
35   The Royal Society's Fellows database does not list a Biographical Memoir.
36   *Biographical Memoirs of Fellows of the Royal Society* Vol 50 2009, pp. 294–5.
37   National Archives AVIA 22/2302.
38   Ibid.
39   National Archives CAB 104/226.
40   Ibid.
41   Ibid.
42   Ibid.
43   National Archives HO 213/587.

Chapter 2

1   *Notes and Records*, The Royal Society, 1946, pp. 1–15.
2   The Right Hon. Admiral of the Fleet, Lord Chatfield GCB OM, Minister for Co-Ordination of Defence.
3   National Archives AVIA 22/2146.
4   Ibid.
5   David Rogers, *Top Secret: British Boffins in World War One* (Solihull: Helion & Company, 2013).
6   National Archives CAB 123/178.
7   Ibid.
8   National Archives CAB 127/218.
9   National Archives CAB 123/178.
10   National Archives CAB 127/214.
11   National Archives CAB 127/218.
12   National Archives CAB 21/1169.
13   National Archives CAB 115/447.
14   Ibid.
15   National Archives CAB 115/447.
16   National Archives CAB 127/218.
17   National Archives CAB 21/830.
18   National Archives INF 1/852.
19   National Archives INF 1/853.
20   *Notes and* Records, The Royal Society, 1946, p. 11.
21   Report dated 29 March 1943, National Archives CO 859/79/7.
22   National Archives DSIR 17/271.

Chapter 3

1   National Archives CAB 90/1.
2   National Archives CAB 90/7.
3   National Archives T161/1288.
4   Ibid.
5   *Who was Who 1951–1960* (London: A. & C. Black Limited, 1967), pp. 1090–1.
6   National Archives T 161/1288.
7   Ibid.
8   National Archives WO 195/1.
9   National Archives AVIA 22/155.
10   National Archives AIR 2/3967.
11   Ibid.
12   The Met Office, *about us* <http://www.metoffice.gov.uk/about-us/who>. (Accessed April 2013).
13   National Archives BJ 5/102.
14   National Archives DSIR 23/12528.
15   *Biographical Memoirs of Fellows of the Royal Society* Vol 17 1971, p. 497.
16   *Biographical Memoirs of Fellows of the Royal Society* Vol 20 1974, p. 487.
17   *Biographical Memoirs of Fellows of the Royal Society* Vol 17 1971, pp. 601–5.
18   *Obituary Notices of Fellows of the Royal Society* Vol 5 1945–1948, pp. 396–7, p. 403.
19   *Biographical Memoirs of Fellows of the Royal Society* Vol 45 1999, p. 38.
20   *Biographical Memoirs of Fellows of the Royal Society* Vol 49 2003, p. 139.
21   *Biographical Memoirs of Fellows of the Royal Society* Vol 9 1963, p. 185.
22   *Biographical Memoirs of Fellows of the Royal Society* Vol 21 1975, pp. 1–115.
23   *Biographical Memoirs of Fellows of the Royal Society* Vol 5 1959, p. 89.
24   *Biographical Memoirs of Fellows of the Royal Society* Vol 17 1971, p. 481.
25   *Biographical Memoirs of Fellows of the Royal Society* Vol 18 1972, p. 543.
26   *Biographical Memoirs of Fellows of the Royal Society* Vol 4 1958, p. 159.
27   *Obituary Notices of Fellows of the Royal Society* Vol 5 1945–1948, p. 153.
28   *Biographical Memoirs of Fellows of the Royal Society* Vol 11 1965, pp. 23–40.
29   *Biographical Memoirs of Fellows of the Royal Society* Vol 17 1971, pp. 224–5.
30   *Biographical Memoirs of Fellows of the Royal Society* Vol 10 1964, p. 223.
31   *Biographical Memoirs of Fellows of the Royal Society* Vol 12 1966, p. 267.
32   *Biographical Memoirs of Fellows of the Royal Society* Vol 17 1971, p. 681.
33   *Biographical Memoirs of Fellows of the Royal Society* Vol 57 2011, pp. 67–8.
34   *Biographical Memoirs of Fellows of the Royal Society* Vol 7 1961, pp. 44–5.

Chapter 4

1.   National Archives FD 1/7042.
2.   National Archives FD 1/7040.
3.   Royal Society, *Collections* <http://royalsociety.org/library/collections/>. (Accessed April 2013).
4.   *Who was Who 1951–1960*, (London: A. & C. Black Limited, 1967), p. 322.
5.   National Archives FD 1/6071.

6.   National Archives FD 1/6066.

7.   National Archives FD 1/6051.

8.   *Who was Who 1971–1980*, (London: A. & C. Black Limited, 1981), pp. 183–4.

9.   National Archives FD 1/6051.

10. National Archives, *Discovery* <http://discovery.nationalarchives.gov.uk/SearchUI/Details?uri=C153>. (Accessed April 2013).

11. *Biographical Memoirs of Fellows of the Royal Society* Vol 46 2000, p. 72.

12. *Biographical Memoirs of Fellows of the Royal Society* Vol 50 2004, p. 6.

13. *Who was Who 1996–2000*, (London: A. & C. Black Limited, 2001), p. 25.

14. *Biographical Memoirs of Fellows of the Royal Society* Vol 45 1999, p. 555.

15. *Biographical Memoirs of Fellows of the Royal Society* Vol 50 2004, pp. 93–107 and Royal Society, *Collections* <http://royalsociety.org/library/collections/>. (Accessed April 2013).

16. *Biographical Memoirs of Fellows of the Royal Society* Vol 54 2008, pp. 171–2.

17. *Biographical Memoirs of Fellows of the Royal Society* Vol 27 1981, pp. 355–64.

18. Charles Mosley (ed), *Burke's Peerage, Baronetage and Knightage: 107th Edition* (London: Burke's Peerage, 2003), p. 2805.

19. *Who was Who, 1971–1980*, (London: A. & C. Black Limited, 1981), p. 565.

20. *Biographical Memoirs of Fellows of the Royal Society* Vol 51 2005, p. 36.

21. *Biographical Memoirs of Fellows of the Royal Society* Vol 54 2008, p. 35.

22. Ibid., p. 50.

23. *Biographical Memoirs of Fellows of the Royal Society* Vol 47 2001, p. 208.

24. *Biographical Memoirs of Fellows of the Royal Society* Vol 48 2002, p. 155.

25. *Biographical Memoirs of Fellows of the Royal Society* Vol 53 2007, pp. 167–8.

26. *Biographical Memoirs of Fellows of the Royal Society* Vol 1 1955, p. 83.

27. *Biographical Memoirs of Fellows of the Royal Society* Vol 45 1999, p. 233.

28. *Biographical Memoirs of Fellows of the Royal Society* Vol 48 2002, p. 226.

29. *Biographical Memoirs of Fellows of the Royal Society* Vol 53 2007, pp. 191–2.

30. *Biographical Memoirs of Fellows of the Royal Society* Vol 45 1999, p. 299.

31. *Biographical Memoirs of Fellows of the Royal Society* Vol 53 2007, p. 239.

32. *Biographical Memoirs of Fellows of the Royal Society* Vol 56 2010, p. 290.

33. *Biographical Memoirs of Fellows of the Royal Society* Vol 27 1981, pp. 379–424.

34. *Biographical Memoirs of Fellows of the Royal Society* Vol 53 2007, pp. 250–1.

35. *Biographical Memoirs of Fellows of the Royal Society* Vol 54 2008, pp. 248–349.

36. *Biographical Memoirs of Fellows of the Royal Society* Vol 56 2010, pp. 363.

37. *Biographical Memoirs of Fellows of the Royal Society* Vol 55 2009, p. 270.

38. *Biographical Memoirs of Fellows of the Royal Society* Vol 45 1999, p. 511.

39. *Biographical Memoirs of Fellows of the Royal Society* Vol 51 2005, p. 54.

40. *Biographical Memoirs of Fellows of the Royal Society* Vol 55 2009, p. 142.

41. *Obituary Notices of Fellows of the Royal Society* Vol 7 1950–1951, p. 162.

42. *Biographical Memoirs of Fellows of the Royal Society* Vol 57 2011, p. 100.

43. *Biographical Memoirs of Fellows of the Royal Society* Vol 46 2000, p. 201.

44. *Biographical Memoirs of Fellows of the Royal Society* Vol 48 2002, p. 408.

45. *Biographical Memoirs of Fellows of the Royal Society* Vol 13 1967, p. 360.

46. *Biographical Memoirs of Fellows of the Royal Society* Vol 45 1999, p. 490.

47. *Biographical Memoirs of Fellows of the Royal Society* Vol 57 2011, p. 210.

48. *Biographical Memoirs of Fellows of the Royal Society* Vol 9 1963, pp. 87–90.
49. Ibid., pp. 173–4.
50. *Biographical Memoirs of Fellows of the Royal Society* Vol 55 2009, p. 81.
51. *Biographical Memoirs of Fellows of the Royal Society* Vol 17 1971, p. 96.
52. *Biographical Memoirs of Fellows of the Royal Society* Vol 52 2006, p. 70.
53. *Biographical Memoirs of Fellows of the Royal Society* Vol 56 2010, p. 66.
54. *Biographical Memoirs of Fellows of the Royal Society* Vol 12 1966, pp. 133-5.
55. *Biographical Memoirs of Fellows of the Royal Society* Vol 51 2005, p. 298.
56. *Biographical Memoirs of Fellows of the Royal Society* Vol 20 1974, p. 350.
57. *Biographical Memoirs of Fellows of the Royal Society* Vol 16 1970, p. 470–1.
58. *Biographical Memoirs of Fellows of the Royal Society* Vol 57 2011, pp. 353–4.
59. Ibid., pp. 298–9.
60. *Biographical Memoirs of Fellows of the Royal Society* Vol 17 1971, p. 146.
61. *Biographical Memoirs of Fellows of the Royal Society* Vol 18 1972, p. 5.
62. Ibid., p. 197.
63. *Biographical Memoirs of Fellows of the Royal Society* Vol 20 1974, p. 143.
64. *Biographical Memoirs of Fellows of the Royal Society* Vol 15 1969, p. 241.
65. *Biographical Memoirs of Fellows of the Royal Society* Vol 18 1972, p. 26.
66. *Biographical Memoirs of Fellows of the Royal Society* Vol 21 1975, p. 209.
67. *Biographical Memoirs of Fellows of the Royal Society* Vol 18 1972, p. 37.
68. *Biographical Memoirs of Fellows of the Royal Society* Vol 13 1967, pp. 193–203.
69. *Biographical Memoirs of Fellows of the Royal Society* Vol 17 1971, p. 744.
70. *Biographical Memoirs of Fellows of the Royal Society* Vol 47 2001, p. 23.
71. *Biographical Memoirs of Fellows of the Royal Society* Vol 16 1970, p. 20.
72. *Biographical Memoirs of Fellows of the Royal Society* Vol 47 2001, p. 41.
73. *Biographical Memoirs of Fellows of the Royal Society* Vol 54 2008, pp. 260–1.
74. *Biographical Memoirs of Fellows of the Royal Society* Vol 19 1973, pp. 335–6.
75. *Biographical Memoirs of Fellows of the Royal Society* Vol 53 2007, pp. 225–6.
76. *Biographical Memoirs of Fellows of the Royal Society* Vol 17 1971, p. 669.
77. Ibid., p. 57.
78. *Biographical Memoirs of Fellows of the Royal Society* Vol 12 1966, p. 343.
79. *Biographical Memoirs of Fellows of the Royal Society* Vol 14 1968, p. 119.
80. *Biographical Memoirs of Fellows of the Royal Society* Vol 52 2006, p. 209.
81. *Biographical Memoirs of Fellows of the Royal Society* Vol 51 2005, p. 225.
82. Nobel Prize, *Physiology or Medicine* (1979) <http://www.nobelprize.org/nobel_prizes/medicine/laureates/1979/hounsfield.html>. (Accessed April 2013).
83. Royal Society, *Collections* <http://royalsociety.org/library/collections/>. (Accessed April 2013).
84. David Rogers, *Nobel Laureate Contributions to 20th Century Chemistry* (London: RSC Publishing, 2006).
85. *Biographical Memoirs of Fellows of the Royal Society* Vol 51 2005, pp. 221-235.

Chapter 5

1. *Biographical Memoirs of Fellows of the Royal Society* Vol 1 1955, p. 7 and *Who was Who 1951–1960*, (London: A. & C. Black Limited, 1967), p. 18.
2. *Biographical Memoirs of Fellows of the Royal Society* Vol 9 1963, p. 31.
3. *Biographical Memoirs of Fellows of the Royal Society* Vol 19 1973, p. 23.
4. *Biographical Memoirs of Fellows of the Royal Society* Vol 5 1959, p. 25.
5. *Biographical Memoirs of Fellows of the Royal Society* Vol 45 1999, p. 24.
6. Royal Society, *Collections* <http://royalsociety.org/library/collections/>. (Accessed April 2013).
7. *Biographical Memoirs of Fellows of the Royal Society* Vol 53 2007, pp. 21–44.
8. *Biographical Memoirs of Fellows of the Royal Society* Vol 19 1973, p. 123.
9. *Biographical Memoirs of Fellows of the Royal Society* Vol 21 1975, p. 169.
10. *Biographical Memoirs of Fellows of the Royal Society* Vol 19 1973, p. 215.
11. *Biographical Memoirs of Fellows of the Royal Society* Vol 46 2000, p. 52.
12. Royal Society, *Collections* <http://royalsociety.org/library/collections/>. (Accessed April 2013).
13. *Biographical Memoirs of Fellows of the Royal Society* Vol 45 1999, p. 153 and Millar, David, Ian, John and Margaret, *The Cambridge Dictionary of Scientists 2nd edition* (Cambridge University Press, 2002), p. 112.
14. *Biographical Memoirs of Fellows of the Royal Society* Vol 51 2005, p. 123.
15. *Biographical Memoirs of Fellows of the Royal Society* Vol 9 1963, pp. 94–5.
16. *Biographical Memoirs of Fellows of the Royal Society* Vol 10 1964, p. 90.
17. *Biographical Memoirs of Fellows of the Royal Society* Vol 10 1964, pp. 109–10.
18. *Biographical Memoirs of Fellows of the Royal Society* Vol 13 1967, pp. 89–106 and Royal Society, *Collections* <http://royalsociety.org/library/collections/>. (Accessed April 2013).
19. *Biographical Memoirs of Fellows of the Royal Society* Vol 47 2001, p. 259 and *Who was Who 1996–2000*, (London: A. & C. Black Limited, 2001), p. 227.
20. *Biographical Memoirs of Fellows of the Royal Society* Vol 12 1966, p. 293.
21. *Biographical Memoirs of Fellows of the Royal Society* Vol 46 2000, p. 253.
22. *Biographical Memoirs of Fellows of the Royal Society* Vol 12 1966, p. 323.
23. *Biographical Memoirs of Fellows of the Royal Society* Vol 10 1964, pp. 149–50.
24. *Biographical Memoirs of Fellows of the Royal Society* Vol 57 2011, pp. 169–77.
25. *Biographical Memoirs of Fellows of the Royal Society* Vol 18 1972, p. 350.
26. Royal Society, *Collections* <http://royalsociety.org/library/collections/>. (Accessed April 2013).
27. *Biographical Memoirs of Fellows of the Royal Society* Vol 20 1974, p. 289.
28. *Biographical Memoirs of Fellows of the Royal Society* Vol 6 1960, pp. 157–68.
29. *Biographical Memoirs of Fellows of the Royal Society* Vol 13 1967, p. 258.
30. *Obituary Notices of Fellows of the Royal Society* Vol 7 1950–1951, p. 240.
31. *Biographical Memoirs of Fellows of the Royal Society* Vol 19 1973, p. 501.
32. *Biographical Memoirs of Fellows of the Royal Society* Vol 48 2002, p. 344.
33. *Biographical Memoirs of Fellows of the Royal Society* vol 17 1971, p. 515.
34. *Obituary Notices of Fellows of the Royal Society* Vol 7 1950–1951, p.256.
35. *Biographical Memoirs of Fellows of the Royal Society* Vol 45 1999, p. 413.

36. *Biographical Memoirs of Fellows of the Royal Society* Vol 51 2005, pp. 367–77.
37. *Biographical Memoirs of Fellows of the Royal Society* Vol 51 2005, p. 381.
38. *Biographical Memoirs of Fellows of the Royal Society* Vol 8 1962, p. 143.
39. *Biographical Memoirs of Fellows of the Royal Society* Vol 10 1964, p. 290 and *Who was Who 1961–70*, (London: A. & C. Black Limited, 1967), p. 1089.
40. *Biographical Memoirs of Fellows of the Royal Society* Vol 52 2006, p. 418.
41. *Biographical Memoirs of Fellows of the Royal Society* Vol 11 1965, p. 182.
42. *Biographical Memoirs of Fellows of the Royal Society* Vol 19 1973, p. 687.
43. *Biographical Memoirs of Fellows of the Royal Society* Vol 10 1964, p. 306.
44. *Biographical Memoirs of Fellows of the Royal Society* Vol 50 2004, pp. 309–13 and *Who was Who 1991–1995*, (London: A. & C. Black Limited, 1996) p. 556.
45. *Who was Who 1961–1970*, (London: A. & C. Black Limited, 1972) p. 1132.
46. *Biographical Memoirs of Fellows of the Royal Society* Vol 8 1962, p. 163.
47. *Biographical Memoirs of Fellows of the Royal Society* Vol 20 1974, pp. 478–479.
48. *Biographical Memoirs of Fellows of the Royal Society* Vol 12 1966, pp. 532–3.
49. *Biographical Memoirs of Fellows of the Royal Society* Vol 2 1956, p. 328 and *Who was Who 1951–1960*, p. 1188.
50. *Biographical Memoirs of Fellows of the Royal Society* Vol 17 1971, p. 619.
51. *Biographical Memoirs of Fellows of the Royal Society* Vol 9 1963, p. 211.
52. *Biographical Memoirs of Fellows of the Royal Society* Vol 17 1971, p. 3.
53. Royal Society Fellows, *Collections* <http://royalsociety.org/library/collections/>. (Accessed April 2013); *Biographical Memoirs of Fellows of the Royal Society* Vol 21 1975, pp. 175–95.
54. *Biographical Memoirs of Fellows of the Royal Society* Vol 46 2000, p. 23.
55. *Biographical Memoirs of Fellows of the Royal Society* Vol 20 1974, p. 80.
56. *Biographical Memoirs of Fellows of the Royal Society* Vol 17 1971, p. 161.
57. *Biographical Memoirs of Fellows of the Royal Society* Vol 19 1973, pp. 283–4.
58. *Biographical Memoirs of Fellows of the Royal Society* Vol 21 1975, pp. 351–3.
59. *Biographical Memoirs of Fellows of the Royal Society* Vol 10 1964, p. 191.
60. *Biographical Memoirs of Fellows of the Royal Society* Vol 21 1975, p. 433.
61. *Biographical Memoirs of Fellows of the Royal Society* Vol 45 1999, p. 318.
62. *Biographical Memoirs of Fellows of the Royal Society* Vol 46 2000, p. 433.
63. *Biographical Memoirs of Fellows of the Royal Society* Vol 11 1965, p. 137.
64. *Biographical Memoirs of Fellows of the Royal Society* Vol 19 1973, p. 566.
65. *Biographical Memoirs of Fellows of the Royal Society* Vol 45 1999, p. 473.
66. *Biographical Memoirs of Fellows of the Royal Society* Vol 49 2003, p. 521–38.
67. *Biographical Memoirs of Fellows of the Royal Society* Vol 52 2006, p. 333.
68. *Biographical Memoirs of Fellows of the Royal Society* Vol 2 1956, p. 300.
69. *Biographical Memoirs of Fellows of the Royal Society* Vol 13 1967, p. 390.
70. *Biographical Memoirs of Fellows of the Royal Society* Vol 46 2000, p. 91.
71. *Biographical Memoirs of Fellows of the Royal Society* Vol 1 1955, pp. 101–17.
72. *Biographical Memoirs of Fellows of the Royal Society* Vol 51 2005, p. 351.
73. *Biographical Memoirs of Fellows of the Royal Society* Vol 17 1971, p. 693.
74. *Biographical Memoirs of Fellows of the Royal Society* Vol 48 2002, pp. 129–50.
75. *Biographical Memoirs of Fellows of the Royal Society* Vol 7 1961, p. 207.
76. *Biographical Memoirs of Fellows of the Royal Society* Vol 17 1971, p. 23.

77. David Rogers, *Nobel Laureate Contributions to 20th Century Chemistry* (London: RSC Publishing, 2006), p. 309.
78. *Biographical Memoirs of Fellows of the Royal Society* Vol 16 1970, p. 210.
79. *Biographical Memoirs of Fellows of the Royal Society* Vol 47 2001, pp. 159–88.
80. Rogers, David. *Nobel Laureate Contributions to 20th Century Chemistry* (London: RSC Publishing, 2006), p. 205.
81. *Biographical Memoirs of Fellows of the Royal Society* Vol 57 2011, p. 133.
82. *Biographical Memoirs of Fellows of the Royal Society* Vol 16 1970, pp. 298–9.
83. *Biographical Memoirs of Fellows of the Royal Society* Vol 49 2003, pp. 179–96.
84. *Biographical Memoirs of Fellows of the Royal Society* Vol 13 1967, p. 173.
85. *Biographical Memoirs of Fellows of the Royal Society* Vol 19 1973, p. 381 and Royal Society, *Collections* <http://royalsociety.org/library/collections/>. (Accessed April 2013).
86. *Biographical Memoirs of Fellows of the Royal Society* Vol 20 1974, p. 251.
87. *Biographical Memoirs of Fellows of the Royal Society* Vol 50 2004, p. 160.
88. *Biographical Memoirs of Fellows of the Royal Society* Vol 14 1968, p. 359.
89. *Biographical Memoirs of Fellows of the Royal Society* Vol 24 1978, p. 454.
90. *Biographical Memoirs of Fellows of the Royal Society* Vol 5 1959, p. 182.
91. *Biographical Memoirs of Fellows of the Royal Society* Vol 17 1971, p. 544.
92. *Biographical Memoirs of Fellows of the Royal Society* Vol 46 2000, p. 448.
93. *Biographical Memoirs of Fellows of the Royal Society* Vol 45 1999, p. 440.
94. *Biographical Memoirs of Fellows of the Royal Society* Vol 17 1971, p. 573.
95. *Biographical Memoirs of Fellows of the Royal Society* Vol 45 1999, pp. 449–67 and Nobel Prize, *Physiology or Medicine* (1950) <http://www.nobelprize.org/nobel_prizes/medicine/laureates/1950/reichstein-facts.html>. (Accessed April 2013).
96. *Biographical Memoirs of Fellows of the Royal Society* Vol 7 1961, p. 224.
97. David Rogers, *Nobel Laureate Contributions to 20th Century Chemistry* (London: RSC Publishing, 2006), p. 127 and *Biographical Memoirs of Fellows of the Royal Society* Vol 18 1972, p. 613.
98. Rogers, David. *Nobel Laureate Contributions to 20th Century Chemistry* (London: RSC Publishing, 2006) and *Biographical Memoirs of Fellows of the Royal Society* Vol 20 1974, p. 414.
99. *Biographical Memoirs of Fellows of the Royal Society* Vol 4 1958, p. 341.
100. *Biographical Memoirs of Fellows of the Royal Society* Vol 45 1999, p. 357.
101. *Biographical Memoirs of Fellows of the Royal Society* Vol 54 2008, pp. 416–17.
102. *Obituary Notices of Fellows of the Royal Society* Vol 9 1954, p. 243.
103. *Biographical Memoirs of Fellows of the Royal Society* Vol 46 2000, p. 537.
104. *Biographical Memoirs of Fellows of the Royal Society* Vol 17 1971, p. 331.
105. *Biographical Memoirs of Fellows of the Royal Society* Vol 19 1973, p. 437.
106. *Biographical Memoirs of Fellows of the Royal Society* Vol 1 1955, p. 1.
107. *Biographical Memoirs of Fellows of the Royal Society* Vol 46 2000, p. 47.
108. *Biographical Memoirs of Fellows of the Royal Society* Vol 47 2001, p. 128.
109. *Biographical Memoirs of Fellows of the Royal Society* Vol 5 1959, p. 49.
110. *Biographical Memoirs of Fellows of the Royal Society* Vol 19 1973, p. 305.
111. *Biographical Memoirs of Fellows of the Royal Society* Vol 5 1959, p. 70.
112. *Biographical Memoirs of Fellows of the Royal Society* Vol 17 1971, pp. 247–8.

113. *Biographical Memoirs of Fellows of the Royal Society* Vol 45 1999, p. 172.

114. *Biographical Memoirs of Fellows of the Royal Society* Vol 45 1999, p. 187.

115. *Biographical Memoirs of Fellows of the Royal Society* Vol 15 1969, pp. 103–4.

116. *Biographical Memoirs of Fellows of the Royal Society* Vol 15 1969, p. 115.

117. *Biographical Memoirs of Fellows of the Royal Society* Vol 18 1972, p. 480.

118. *Biographical Memoirs of Fellows of the Royal Society* Vol 11 1965, pp. 171–3.

119. *Biographical Memoirs of Fellows of the Royal Society* Vol 18 1972, p. 578.

120. *Biographical Memoirs of Fellows of the Royal Society* Vol 7 1961, p. 276.

121. Royal Society, *Collections* <http://royalsociety.org/library/collections/>. (Accessed April 2013) and *Biographical Memoirs of Fellows of the Royal Society* Vol 17 1971, pp. 713–40.

122. *Biographical Memoirs of Fellows of the Royal Society* Vol 47 2001, pp. 497–514.

123. *Biographical Memoirs of Fellows of the Royal Society* Vol 53 2007, p. 406.

124. *Biographical Memoirs of Fellows of the Royal Society* Vol 13 1967, p. 61.

125. *Biographical Memoirs of Fellows of the Royal Society* Vol 52 2006, p. 4.

126. *Biographical Memoirs of Fellows of the Royal Society* Vol 8 1962, p. 7.

127. *Biographical Memoirs of Fellows of the Royal Society* Vol 15 1969, p. 13.

128. *Biographical Memoirs of Fellows of the Royal Society* Vol 50 2004 pp. 43–4.

129. *Biographical Memoirs of Fellows of the Royal Society* Vol 21 1975, p. 277.

130. *Biographical Memoirs of Fellows of the Royal Society* Vol 9 1963, p. 141.

131. *Biographical Memoirs of Fellows of the Royal Society* Vol 46 2000, p. 9.

132. *Biographical Memoirs of Fellows of the Royal Society* Vol 16 1970, pp. 5–6.

133. *Biographical Memoirs of Fellows of the Royal Society* Vol 45 1999, p. 139.

134. *Biographical Memoirs of Fellows of the Royal Society* Vol 53 2007, p. 147.

135. *Biographical Memoirs of Fellows of the Royal Society* Vol 16 1970, p. 344.

136. *Biographical Memoirs of Fellows of the Royal Society* Vol 8 1962, p. 82.

137. *Biographical Memoirs of Fellows of the Royal Society* Vol 56 2010, pp. 257–72.

138. *Biographical Memoirs of Fellows of the Royal Society* Vol 19 1973, p. 225 and Royal Society, *Collections* <http://royalsociety.org/library/collections/>. (Accessed April 2013).

139. *Biographical Memoirs of Fellows of the Royal Society* Vol 12 1966, p. 480.

140. *Biographical Memoirs of Fellows of the Royal Society* Vol 47 2001, p. 484.

141. *Biographical Memoirs of Fellows of the Royal Society* Vol 51 2005, pp. 171–3.

142. *Biographical Memoirs of Fellows of the Royal Society* Vol 45 1999, p. 538.

143. *Biographical Memoirs of Fellows of the Royal Society* Vol 17 1971, pp. 400–3.

144. *Biographical Memoirs of Fellows of the Royal Society* Vol 15 1969, p. 195.

145. *Biographical Memoirs of Fellows of the Royal Society* Vol 47 2001, p. 458.

146. *Biographical Memoirs of Fellows of the Royal Society* Vol 15 1969, p. 91.

147. *Biographical Memoirs of Fellows of the Royal Society* Vol 1 1955, p. 166.

148. *Biographical Memoirs of Fellows of the Royal Society* Vol 16 1970, p. 321.

149. *Biographical Memoirs of Fellows of the Royal Society* Vol 49 2003, p. 512.

150. *Biographical Memoirs of Fellows of the Royal Society* Vol 51 2005, p. 103.

151. *Biographical Memoirs of Fellows of the Royal Society* Vol 19 1973, p. 135.

152. *Biographical Memoirs of Fellows of the Royal Society* Vol 51 2005 pp. 69–71.

153. *Biographical Memoirs of Fellows of the Royal Society* Vol 11 1965, p. 10–11.

154. *Biographical Memoirs of Fellows of the Royal Society* Vol 49 2003, p. 422.

155. *Biographical Memoirs of Fellows of the Royal Society* Vol 16 1970, p. 490.
156. *Biographical Memoirs of Fellows of the Royal Society* Vol 16 1970, p. 259.
157. *Biographical Memoirs of Fellows of the Royal Society* Vol 14 1968, p. 57.
158. *Biographical Memoirs of Fellows of the Royal Society* Vol 2 1956, pp. 291–8.

Chapter 6

1. *Biographical Memoirs of Fellows of the Royal Society* Vol 16 1970, pp. 71–2 and *Who was Who 1961–1970*, (London: A. & C. Black Limited, 1972), p. 270.
2. National Archives HO 217/2.
3. National Archives HO 217/1.
4. Bletchley Park, *History – Wartime History* <http://www.bletchleypark.org.uk/content/hist/early.rhtm>. (Accessed April 2013).
5. Bletchley Park, *Roll of Honour* <http://rollofhonour.bletchleypark.org.uk/search/>. (Accessed April 2013).
6. Alan Turing, *A short biography by Andrew Hodges* <http://www.turing.org.uk/bio/part4.html>. (Accessed April 2013).
7. National Archives HW 25/1.
8. *Biographical Memoirs of Fellows of the Royal Society* Vol 1 1955, p. 254.
9. National Archives HW 25/1.
10. Bletchley Park, *History – Wartime History* <http://www.bletchleypark.org.uk/content/hist/early.rhtm>. (Accessed April 2013).
11. Alan Turing, *A short biography by Andrew Hodges* <http://www.turing.org.uk/bio/part4.html>. (Accessed April 2013).
12. Bletchley Park, *History – Wartime History* <http://www.bletchleypark.org.uk/content/hist/early.rhtm> (Accessed April 2013).
13. *Biographical Memoirs of Fellows of the Royal Society* Vol 31 1985, pp. 435–52.
14. National Archives HW 25/24.
15. Ibid.
16. *Biographical Memoirs of Fellows of the Royal Society* Vol 55 2009, pp. 260–1.
17. *Biographical Memoirs of Fellows of the Royal Society* Vol 52 2006, p. 319.
18. *Biographical Memoirs of Fellows of the Royal Society* Vol 45 1999, p. 205.
19. *Biographical Memoirs of Fellows of the Royal Society* Vol 19 1973, p. 74.
20. *Biographical Memoirs of Fellows of the Royal Society* Vol 9 1963, p. 297.
21. *Biographical Memoirs of Fellows of the Royal Society* Vol 48 2002, p. 4 and Royal Society Fellows, *Collections* <http://royalsociety.org/library/collections/>. (Accessed April 2013).
22. *Biographical Memoirs of Fellows of the Royal Society* Vol 46 2000, p. 182.
23. Royal Society, *Collections* <http://royalsociety.org/library/collections/>. (Accessed April 2013) and *Biographical Memoirs of Fellows of the Royal Society* Vol 39 1994, p. 368.
24. *Biographical Memoirs of Fellows of the Royal Society* Vol 45 1999, p. 243; The Royal Society of Edinburgh, *Obituaries* <http://www.rse.org.uk/620_ObituariesJ.html>. (Accessed April 2013) and *The Independent*, Friday 19 December 1997 <http://www.independent.co.uk/news/obituaries/obituary-professor-r-v-jones-1289581.html>. (Accessed April 2013).

25. *Biographical Memoirs of Fellows of the Royal Society* Vol 52 2006, p. 288.
26. National Archives AVIA 7/964.
27. National Archives AVIA 7/837.
28. *Biographical Memoirs of Fellows of the Royal Society* Vol 49 2003, p. 4.
29. *Biographical Memoirs of Fellows of the Royal Society* Vol 53 2007, p. 114 and Royal Society, *Collections* <http://royalsociety.org/library/collections/>. (Accessed April 2013).
30. *Biographical Memoirs of Fellows of the Royal Society* Vol 49 2003, p. 111.
31. *Biographical Memoirs of Fellows of the Royal Society* Vol 47 2001, pp. 71–2.
32. *Biographical Memoirs of Fellows of the Royal Society* Vol 14 1968, pp. 153–60.
33. *Biographical Memoirs of Fellows of the Royal Society* Vol 45 1999, p. 99.
34. *Biographical Memoirs of Fellows of the Royal Society* Vol 48 2002, pp. 170–1.
35. *Biographical Memoirs of Fellows of the Royal Society* Vol 49 2003, p. 202.
36. See Hey's Biographical Memoir.
37. *Biographical Memoirs of Fellows of the Royal Society* Vol 17 1971, p. 383.
38. *Biographical Memoirs of Fellows of the Royal Society* Vol 47 2001, p. 314.
39. *Biographical Memoirs of Fellows of the Royal Society* Vol 48 2002, p. 293.
40. See Hey's Biographical Memoir.
41. *Biographical Memoirs of Fellows of the Royal Society* Vol 51 2009, p. 205.
42. *Biographical Memoirs of Fellows of the Royal Society* Vol 19 1973, p. 640.
43. See Curran's Biographical Memoir.
44. See Curran's Biographical Memoir.
45. *Biographical Memoirs of Fellows of the Royal Society* Vol 6 1960, p. 262 and Curran's Biographical Memoir.
46. See Hey's Biographical Memoir.
47. See Hey's Biographical Memoir.
48. *Biographical Memoirs of Fellows of the Royal Society* Vol 51 2009, p. 5.
49. *Biographical Memoirs of Fellows of the Royal Society* Vol 50 2004, p. 260.
50. *Biographical Memoirs of Fellows of the Royal Society* Vol 49 2003, p. 123.
51. *Biographical Memoirs of Fellows of the Royal Society* Vol 51 2005, p. 360.
52. Ibid., p. 90.
53. *Biographical Memoirs of Fellows of the Royal Society* Vol 49 2003, p. 217.
54. *Biographical Memoirs of Fellows of the Royal Society* Vol 54 2008, p. 196.
55. *Biographical Memoirs of Fellows of the Royal Society* Vol 48 2002, p. 393.
56. *Biographical Memoirs of Fellows of the Royal Society* Vol 16 1970, pp. 445–6.
57. *Biographical Memoirs of Fellows of the Royal Society* Vol 52 2006, p. 120.
58. *Biographical Memoirs of Fellows of the Royal Society* Vol 10 1964, pp. 232–4.
59. *Biographical Memoirs of Fellows of the Royal Society* Vol 53 2007, pp. 191–192.
60. Ibid., pp. 1–20.
61. *Biographical Memoirs of Fellows of the Royal Society* Vol 12 1966, pp. 148–9.
62. *Biographical Memoirs of Fellows of the Royal Society* Vol 53 2007, p. 341.
63. *Biographical Memoirs of Fellows of the Royal Society* Vol 54 2008, p. 12.
64. *Biographical Memoirs of Fellows of the Royal Society* Vol 4 1958, p. 177.
65. *Biographical Memoirs of Fellows of the Royal Society* Vol 8 1962, pp. 109–13.
66. National Archives WO 188/1809.

Chapter 7

1. National Archives AVIA 22/2305.
2. *Who was Who 1951–1960*, (London: A. & C. Black Limited, 1967), p. 464.
3. National Archives AVIA 22/2305.
4. National Archives AVIA 22/848.
5. National Archive SUPP 6/951.
6. Further information on the *Science of Explosions*, a series of monographs complied in 1956 but concerning 1939–45 was written by Cecil Edwin Henry Bawn FRS and Godfrey Rotter, National Archives SUPP 22/33.
7. National Archives WO 188/1784.
8. *Biographical Memoirs of Fellows of the Royal Society* Vol 35 1990, pp. 381–402.
9. *Biographical Memoirs of Fellows of the Royal Society* Vol 17 1971, pp. 307–9.
10. *Biographical Memoirs of Fellows of the Royal Society* Vol 50 2004, p. 24.
11. *Biographical Memoirs of Fellows of the Royal Society* Vol 43 1997, p. 51.
12. *Biographical Memoirs of Fellows of the Royal Society* Vol 19 1973, pp. 97–8.
13. *Biographical Memoirs of Fellows of the Royal Society* Vol 50 2004, p. 52.
14. *Biographical Memoirs of Fellows of the Royal Society* Vol 50 2004, p. 64.
15. *Biographical Memoirs of Fellows of the Royal Society* Vol 50 2004, p. 150.
16. *Biographical Memoirs of Fellows of the Royal Society* Vol 55 2009, pp. 111, 115.
17. *Biographical Memoirs of Fellows of the Royal Society* Vol 46 2000, p. 274.
18. *Biographical Memoirs of Fellows of the Royal Society* Vol 55 2009, p. 125.
19. *Obituary Notices of Fellows of the Royal Society* Vol 7 1950–1951, p. 424.
20. *Biographical Memoirs of Fellows of the Royal Society* Vol 57 2011, p. 275.
21. *Biographical Memoirs of Fellows of the Royal Society* Vol 51 2005, pp. 305, 308.
22. *Biographical Memoirs of Fellows of the Royal Society* Vol 48 2002, p. 421.
23. *Obituary Notices of Fellows of the Royal Society* Vol 7 1950–1951, p. 495.
24. *Biographical Memoirs of Fellows of the Royal Society* Vol 9 1963, p. 308.
25. *Biographical Memoirs of Fellows of the Royal Society* Vol 49 2003, pp. 541–2.
26. National Archives AB 16/226.
27. National Archives CAB 126/329.
28. National Archives AB 1/714.
29. National Archives CAB 126/46.
30. Presidents of the Royal Society use the post nominal letters PRS denoting President of the Royal Society.
31. National Archives AB 1/206.
32. National Archives AB 1/215.
33. National Archives AB 16/215B.
34. *Biographical Memoirs of Fellows of the Royal Society* Vol 48 2002, p. 90.
35. *Biographical Memoirs of Fellows of the Royal Society* Vol 10 1964, p. 259.
36. *Biographical Memoirs of Fellows of the Royal Society* Vol 6 1960, p. 229.
37. *Biographical Memoirs of Fellows of the Royal Society* Vol 6 1960, p. 262.
38. *Biographical Memoirs of Fellows of the Royal Society* Vol 46 2000, p. 306.
39. *Biographical Memoirs of Fellows of the Royal Society* Vol 9 1963, pp. 45–47.
40. *Biographical Memoirs of Fellows of the Royal Society* Vol 1 1955, p. 73.
41. *Biographical Memoirs of Fellows of the Royal Society* Vol 46 2000, p. 587.

42. *Biographical Memoirs of Fellows of the Royal Society* Vol 53 2007, p. 332.
43. *Biographical Memoirs of Fellows of the Royal Society* Vol 29 1983, pp. 623, 641, 643.
44. *Biographical Memoirs of Fellows of the Royal Society* Vol 14 1968, p. 399.
45. *Biographical Memoirs of Fellows of the Royal Society* Vol 11 1965, p. 64.
46. *Biographical Memoirs of Fellows of the Royal Society* Vol 52 2006, p. 460.
47. *Biographical Memoirs of Fellows of the Royal Society* Vol 48 2002, p. 106.
48. *Biographical Memoirs of Fellows of the Royal Society* Vol 35 1990, p. 334.
49. *Biographical Memoirs of Fellows of the Royal Society* Vol 47 2001, p. 390.
50. National Archives AIR 20/11994.
51. National Archives AVIA 15/700.
52. National Archives AVIA 15/744.
53. Royal Society, *Collections* <http://royalsociety.org/library/collections/>. (Accessed April 2013).
54. National Archives AVIA 15/744.
55. National Archives AVIA 15/3933.
56. National Archives AIR 14/844.
57. *Biographical Memoirs of Fellows of the Royal Society* Vol 43 1997, p. 5.
58. *Biographical Memoirs of Fellows of the Royal Society* Vol 10 1964, p. 6.
59. *Biographical Memoirs of Fellows of the Royal Society* Vol 14 1968, pp. 89-90.
60. *Biographical Memoirs of Fellows of the Royal Society* Vol 16 1970, pp. 241.
61. *Biographical Memoirs of Fellows of the Royal Society* Vol 50 2004, pp. 96-7.
62. *Biographical Memoirs of Fellows of the Royal Society* Vol 46 2000, pp. 177-96.
63. *Obituary Notices of Fellows of the Royal Society* Vol 7 1950–1951, p. 118.
64. *Biographical Memoirs of Fellows of the Royal Society* Vol 12 1966, p. 167.
65. *Biographical Memoirs of Fellows of the Royal Society* Vol 16 1970, pp. 334–5.
66. *Biographical Memoirs of Fellows of the Royal Society* Vol 49 2003, p. 266.
67. *Biographical Memoirs of Fellows of the Royal Society* Vol 54 2008, p. 381.
68. *Biographical Memoirs of Fellows of the Royal Society* Vol 12 1966, p. 551.

Chapter 8

1   National Archives AIR 2/1933.
2   National Archives AIR 14/3604.
3   National Archives DSIR 23/7778.
4   National Archives ADM 116/4234.
5   Ibid.
6   National Archives AIR 62/800.
7   National Archives AVIA 15/2406.
8   *Biographical Memoirs of Fellows of the Royal Society* Vol 10 1964, p. 125.
9   *Biographical Memoirs of Fellows of the Royal Society* Vol 51 2005, p. 139.
10   *Biographical Memoirs of Fellows of the Royal Society* Vol 19 1973, pp. 273–6.
11   *Biographical Memoirs of Fellows of the Royal Society* Vol 8 1962, pp. 125–6.
12   National Archives AVIA 15/2406.
13   *Biographical Memoirs of Fellows of the Royal Society* Vol 20 1974, pp. 239–40.
14   *Biographical Memoirs of Fellows of the Royal Society* Vol 19 1973, pp. 359–62.

15   *Biographical Memoirs of Fellows of the Royal Society* Vol 45 1999, pp. 423–4.
16   *Biographical Memoirs of Fellows of the Royal Society* Vol 16 1970, p. 551.
17   *Biographical Memoirs of Fellows of the Royal Society* Vol 1 1955, pp. 163–73.
18   *Biographical Memoirs of Fellows of the Royal Society* Vol 16 1970, p. 210.
19   *Biographical Memoirs of Fellows of the Royal Society* Vol 50 2004, p. 288.
20   *Biographical Memoirs of Fellows of the Royal Society* Vol 47 2001, pp. 453–64.
21   *Biographical Memoirs of Fellows of the Royal Society* Vol 49 2003, p. 151.
22   *Biographical Memoirs of Fellows of the Royal Society* Vol 18 1972, pp. 611, 613.
23   *Biographical Memoirs of Fellows of the Royal Society* Vol 16 1970, p. 377.
24   National Archives BT 200/1.
25   *Biographical Memoirs of Fellows of the Royal Society* Vol 10 1964, p. 28.
26   National Archives DSIR 36/1382.
27   *Biographical Memoirs of Fellows of the Royal Society* Vol 46 2000, pp. 125–43.
28   *Biographical Memoirs of Fellows of the Royal Society* Vol 51 2005, pp. 197–9.
29   *Biographical Memoirs of Fellows of the Royal Society* Vol 47 2001, p. 446.
30   *Biographical Memoirs of Fellows of the Royal Society* Vol 51 2005, p. 256.
31   *Biographical Memoirs of Fellows of the Royal Society* Vol 47 2001, p. 420.
32   *Biographical Memoirs of Fellows of the Royal Society* Vol 49 2003, pp. 434–5.
33   *Biographical Memoirs of Fellows of the Royal Society* Vol 52 2006, p. 234.
34   *Biographical Memoirs of Fellows of the Royal Society* Vol 50 2004, p. 80.
35   *Biographical Memoirs of Fellows of the Royal Society* Vol 47 2001, pp. 239–53.
36   *Biographical Memoirs of Fellows of the Royal Society* Vol 51 2005, pp. 305, 308.
37   *Biographical Memoirs of Fellows of the Royal Society* Vol 54 2008, p. 434.
38   *Biographical Memoirs of Fellows of the Royal Society* Vol 49 2003, p. 371.
39   *Biographical Memoirs of Fellows of the Royal Society* Vol 55 2009, p. 95.
40   *Biographical Memoirs of Fellows of the Royal Society* Vol 47 2001, p. 228.
41   Ibid., pp. 282–3.
42   *Biographical Memoirs of Fellows of the Royal Society* Vol 48 2002, p. 28.
43   *Biographical Memoirs of Fellows of the Royal Society* Vol 50 2004, p. 318.
44   *Biographical Memoirs of Fellows of the Royal Society* Vol 12 1966, p. 499.

Chapter 9

1.   National Archives MAF 33/417.
2.   Ibid.
3.   National Archives CAB 21/1169.
4.   National Archives DSIR 13/601.
5.   Rothamsted Research, *About-Us* <http://www.rothamsted.ac.uk>. (Accessed April 2013).
6.   *Biographical Memoirs of Fellows of the Royal Society* Vol 12 1966, p. 221.
7.   *Biographical Memoirs of Fellows of the Royal Society* Vol 12 1966, p. 254 and East Malling Research <http://www.emr.ac.uk/>. (Accessed April 2013).
8.   *Biographical Memoirs of Fellows of the Royal Society* Vol 47 2001, p. 192.
9.   *Obituary Notices of Fellows of the Royal Society* Vol 7 1950–1951, pp. 96, 98.
10.   *Biographical Memoirs of Fellows of the Royal Society* Vol 21 1975, p. 410.
11.   *Biographical Memoirs of Fellows of the Royal Society* Vol 13 1967, p. 275.

12. *Biographical Memoirs of Fellows of the Royal Society* Vol 47 2001, p. 373.
13. *Biographical Memoirs of Fellows of the Royal Society* Vol 1 1972, p. 65.
14. *Biographical Memoirs of Fellows of the Royal Society* Vol 49 2003, p. 402.
15. *Biographical Memoirs of Fellows of the Royal Society* Vol 12 1966, p. 470.
16. *Biographical Memoirs of Fellows of the Royal Society* Vol 1 1955, p. 242.
17. *Biographical Memoirs of Fellows of the Royal Society* Vol 49 2003, p. 453.
18. *Biographical Memoirs of Fellows of the Royal Society* Vol 48 2002, p. 444.
19. *Biographical Memoirs of Fellows of the Royal Society* Vol 12 1966, p. 508.
20. *Obituary Notices of Fellows of the Royal Society* Vol 8 1952-1953, p. 311.
21. *Biographical Memoirs of Fellows of the Royal Society* Vol 53 2007, pp. 80–1.
22. *Biographical Memoirs of Fellows of the Royal Society* Vol 55 2009, p. 47.
23. *Biographical Memoirs of Fellows of the Royal Society* Vol 11 1965, pp. 105–6.
24. *Biographical Memoirs of Fellows of the Royal Society* Vol 50 2004, p. 137.
25. *Biographical Memoirs of Fellows of the Royal Society* Vol 46 2000, p. 486.
26. *Biographical Memoirs of Fellows of the Royal Society* Vol 53 2007, pp. 290–2.
27. *Biographical Memoirs of Fellows of the Royal Society* Vol 48 2002, p. 360.
28. National Archives FD 1/7043.
29. *Biographical Memoirs of Fellows of the Royal Society* Vol 48 2002, p. 240.
30. *Biographical Memoirs of Fellows of the Royal Society* Vol 47 2001, p. 519.
31. *Biographical Memoirs of Fellows of the Royal Society* Vol 49 2003, p. 168.
32. Ibid., p. 464.
33. *Biographical Memoirs of Fellows of the Royal Society* Vol 2 1956, p. 123.
34. *Biographical Memoirs of Fellows of the Royal Society* Vol 2 1956, p. 74.
35. *Biographical Memoirs of Fellows of the Royal Society* Vol 48 2002, p. 465.
36. *Biographical Memoirs of Fellows of the Royal Society* Vol 12 1966, p. 313.
37. *Biographical Memoirs of Fellows of the Royal Society* Vol 10 1964, p. 63.
38. *Biographical Memoirs of Fellows of the Royal Society* Vol 48 2002, p. 492.
39. *Biographical Memoirs of Fellows of the Royal Society* Vol 1 1955, p. 201.
40. *Biographical Memoirs of Fellows of the Royal Society* Vol 10 1964, p. 42.
41. *Biographical Memoirs of Fellows of the Royal Society* Vol 48 2002, pp. 54–5.
42. National Archives, DSIR 5/31.
43. Nobel Prize, Physiology or Medicine (1945) <http://www.nobelprize.org/nobel_prizes/medicine/>. (Accessed April 2013).
44. *Biographical Memoirs of Fellows of the Royal Society* Vol 46 2000, p. 522.
45. *Biographical Memoirs of Fellows of the Royal Society* Vol 22 1976, p. 425.
46. *Biographical Memoirs of Fellows of the Royal Society* Vol 27 1981, p. 634.
47. *Biographical Memoirs of Fellows of the Royal Society* Vol 17 1971, p. 267.
48. *Biographical Memoirs of Fellows of the Royal Society* Vol 48 2002, p. 189.
49. Georgina Ferry, *Dorothy Hodgkin A Life* (Granta Books, 1998).
50. Ibid, p. 89.
51. *Biographical Memoirs of Fellows of the Royal Society* Vol 18 1972, p. 496.
52. *Biographical Memoirs of Fellows of the Royal Society* Vol 46 2000, p. 468.
53. *Biographical Memoirs of Fellows of the Royal Society* Vol 49 2003, p. 19.
54. *Biographical Memoirs of Fellows of the Royal Society* Vol 11 1965, pp. 206–7.
55. *Biographical Memoirs of Fellows of the Royal Society* Vol 51 2005, p.298.
56. *Biographical Memoirs of Fellows of the Royal Society* Vol 56 2010, pp. 4–23.

57. *Biographical Memoirs of Fellows of the Royal Society* Vol 42 1996, pp. 315–38.
58. *Biographical Memoirs of Fellows of the Royal Society* Vol 20 1974, p. 297.
59. *Biographical Memoirs of Fellows of the Royal Society* Vol 16 1970, pp. 445–6.
60. *Biographical Memoirs of Fellows of the Royal Society* Vol 13 1967, p. 144.
61. *Biographical Memoirs of Fellows of the Royal Society* Vol 18 1972, p. 232.
62. *Obituary Notices of Fellows of the Royal Society* Vol 8 1952–1953, p. 153.

Chapter 10

1 *Notes and Records*, The Royal Society, 1946, p. 4.
2 Royal Society AE 1/9/1.
3 National Archives ADM 1/19723.
4 National Archives BW 2/335.
5 National Archives DSIR 17/271.

Appendix I

1 National Archives, CAB 123/178.
2 Except the Admiralty Research Laboratory all these Establishments are under the executive control of Naval Officers.

Appendix II

1 National Archives, CAB 90/1.

Appendix III

1 National Archives, CAB 90/1.

Appendix IV

1 National Archives, AVIA 22/178.

Appendix V

1 National Archives, CAB 21/830.

Appendix VI

1 National Archives, AIR 2/3967.

Appendix VII

1 National Archives, HW 64/68.

Appendix VIII

1   National Archives, AVIA 22/2305.

Appendix IX

1   National Archives, AB 16/226.

Appendix X

1   National Archives, AIR 14/844.

Appendix XI

1   National Archives DSIR, 23/7778.

Bibliography

Archive documents

National Archives:
AB 1/206, AB 1/215, AB 1/714, AB 16/215B, AB 16/226, ADM 1/19723, ADM 116/4234, AIR 14/3604, AIR 14/844, AIR 2/1933, AIR 2/3967, AIR 20/11994, AIR 62/800, AVIA 15/2406, AVIA 15/3933, AVIA 15/700, AVIA 15/744, AVIA 22/155, AVIA 22/178, AVIA 22/2146, AVIA 22/2146, AVIA 22/2302, AVIA 22/2305, AVIA 22/848, AVIA 7/837, AVIA 7/964, BJ 5/102, BT 200/1, BW 2/335, CAB 90/1, CAB 90/2CAB 104/226, CAB 115/447, CAB 123/178, CAB 126/329, CAB 126/46, CAB 127/214, CAB 127/218, CAB 21/1169, CAB 21/830, CAB 90/1, CAB 90/7, CO 859/79/7, DSIR 13/601, DSIR 17/271, DSIR 23/12528, DSIR 23/7778, DSIR 23/7778, DSIR 36/1382, DSIR 5/31, FD 1/6051, FD 1/6066, FD 1/6071, FD 1/7040, FD 1/7042, FD 1/7043, HO 213/587, HO 217/1, HO 217/2, HW 25/1, HW 25/1, HW 25/24, HW 25/31, HW 64/68, HW 77/7, INF 1/852, INF 1/853, MAF 33/417, SUPP 6/951, SUPP 22/33, T 161/1288, T161/1288, WO 188/1784, WO 188/1809, WO 195/1.

Royal Society Archives:
AE1/9/1.

Printed Books

Brück, Hermann A. *The Story of Astronomy in Edinburgh* (Edinburgh: Edinburgh University Press, 1983).

David, Ian, John and Margaret Millar, *The Cambridge Dictionary of Scientists: 2nd Edition* (Cambridge: Cambridge University Press, 2002).

Ferry, Georgina, *Dorothy Hodgkin A Life* (London: Granta Books, 1998).

Mosley, Charles (ed.), *Burke's Peerage, Baronetage and Knightage: 107th Edition* (London: Burke's Peerage, 2003).

Rogers, David, *Nobel Laureate Contributions to 20th Century Chemistry* (London: RSC Publishing, 2006).

—— *Top Secret: British Boffins in World War One* (Solihull, Helion & Company Ltd, 2013).

Whitaker, Joseph, *Whitaker's Almanack 1941*(Whitaker, 1941).

Who was Who 1951–1960, (London: A. & C. Black Limited, 1967).

Who was Who 1961–1970, (London: A. & C. Black Limited, 1972).

Who was Who 1971–1980, (London: A. & C. Black Limited, 1981).

Who was Who 1991–1995, (London: A. & C. Black Limited, 1996).

Who was Who 1996–2000, (London: A. & C. Black Limited, 2001).

Royal Society Publications

Biographical Memoirs of Fellows of the Royal Society Vol 1 1955.
Biographical Memoirs of Fellows of the Royal Society Vol 2 1956.
Biographical Memoirs of Fellows of the Royal Society Vol 4 1958.
Biographical Memoirs of Fellows of the Royal Society Vol 5 1959.
Biographical Memoirs of Fellows of the Royal Society Vol 6 1960.
Biographical Memoirs of Fellows of the Royal Society Vol 7 1961.
Biographical Memoirs of Fellows of the Royal Society Vol 8 1962.
Biographical Memoirs of Fellows of the Royal Society Vol 9 1963.
Biographical Memoirs of Fellows of the Royal Society Vol 10 1964.
Biographical Memoirs of Fellows of the Royal Society Vol 11 1965.
Biographical Memoirs of Fellows of the Royal Society Vol 12 1966.
Biographical Memoirs of Fellows of the Royal Society Vol 13 1967.
Biographical Memoirs of Fellows of the Royal Society Vol 14 1968.
Biographical Memoirs of Fellows of the Royal Society Vol 15 1969.
Biographical Memoirs of Fellows of the Royal Society Vol 16 1970.
Biographical Memoirs of Fellows of the Royal Society Vol 17 1971.
Biographical Memoirs of Fellows of the Royal Society Vol 18 1972.
Biographical Memoirs of Fellows of the Royal Society Vol 19 1973.
Biographical Memoirs of Fellows of the Royal Society Vol 20 1974.
Biographical Memoirs of Fellows of the Royal Society Vol 21 1975.
Biographical Memoirs of Fellows of the Royal Society Vol 22 1976.
Biographical Memoirs of Fellows of the Royal Society Vol 24 1978.
Biographical Memoirs of Fellows of the Royal Society Vol 27 1981.
Biographical Memoirs of Fellows of the Royal Society Vol 29 1983.
Biographical Memoirs of Fellows of the Royal Society Vol 31 1985.
Biographical Memoirs of Fellows of the Royal Society Vol 35 1990.
Biographical Memoirs of Fellows of the Royal Society Vol 39 1994.
Biographical Memoirs of Fellows of the Royal Society Vol 42 1996.
Biographical Memoirs of Fellows of the Royal Society Vol 43 1997.
Biographical Memoirs of Fellows of the Royal Society Vol 45 1999.
Biographical Memoirs of Fellows of the Royal Society Vol 46 2000.

Biographical Memoirs of Fellows of the Royal Society Vol 47 2001.
Biographical Memoirs of Fellows of the Royal Society Vol 48 2002.
Biographical Memoirs of Fellows of the Royal Society Vol 49 2003.
Biographical Memoirs of Fellows of the Royal Society Vol 50 2004.
Biographical Memoirs of Fellows of the Royal Society Vol 51 2005.
Biographical Memoirs of Fellows of the Royal Society Vol 52 2006.
Biographical Memoirs of Fellows of the Royal Society Vol 53 2007.
Biographical Memoirs of Fellows of the Royal Society Vol 54 2008.
Biographical Memoirs of Fellows of the Royal Society Vol 55 2009.
Biographical Memoirs of Fellows of the Royal Society Vol 56 2010.
Biographical Memoirs of Fellows of the Royal Society Vol 57 2011.
Notes and Records, The Royal Society, 1940–41.
Notes and Records, The Royal Society, 1946.
Obituary Notices of Fellows of the Royal Society 1945–1948 Vol 5.
Obituary Notices of Fellows of the Royal Society 1950–1951 Vol 7.
Obituary Notices of Fellows of the Royal Society 1952–1953 Vol 8.
Obituary Notices of Fellows of the Royal Society 1954 Vol 9.

Websites

Alan Turing, *A short biography by Andrew Hodges* <http://www.turing.org.uk/bio/part4.html>. (Accessed April 2013).

Bletchley Park, *History – Wartime History* <http://www.bletchleypark.org.uk/content/hist/early.rhtm>. (Accessed April 2013).

Bletchley Park, *Roll of Honour* <http://rollofhonour.bletchleypark.org.uk/search/>. (Accessed April 2013).

Bodleian Library University of Oxford, *Catalogue of the Archive of the Society for the Protection of Science and Learning* (1933–87) <http://www.bodley.ox.ac.uk/dept/scwmss/wmss/online/modern/spsl/spsl.html>. (Accessed April 2013).

Council for Assisting Refugee Academics, *Academic Freedom* <http://www.academic-refugees.org/>. (Accessed April 2013).

East Malling Research <http://www.emr.ac.uk/>. (Accessed April 2013).

National Archives, *Discovery* <http://discovery.nationalarchives.gov.uk/SearchUI/Details?uri=C153>. (Accessed April 2013).

Nobel Prize, *Physiology or Medicine* (1945) < http://www.nobelprize.org/nobel_prizes/medicine/>. (Accessed April 2013).

Nobel Prize, *Physiology or Medicine* (1950) <http://www.nobelprize.org/nobel_prizes/medicine/laureates/1950/reichstein-facts.html>. (Accessed April 2013).

Nobel Prize, *Physiology or Medicine* (1979) <http://www.nobelprize.org/nobel_prizes/medicine/laureates/1979/hounsfield.html>. (Accessed April 2013).

Rothamsted Research, *About-Us* <http://www.rothamsted.ac.uk>. (Accessed April 2013).

Royal Society, *Collections* <http://royalsociety.org/library/collections/>. (Accessed April 2013).

The Independent, Friday 19 December 1997 <http://www.independent.co.uk/news/obituaries/obituary-professor-r-v-jones-1289581.html>. (Accessed April 2013).

The Met Office, *about us* <http://www.metoffice.gov.uk/about-us/who>. (Accessed April 2013).

The Royal Society of Edinburgh, *Obituaries* <http://www.rse.org.uk/620_ObituariesJ.html>. (Accessed April 2013)

The Royal Society, *Officers* <http://royalsociety.org/about-us/governance/officers/>. (Accessed April 2013).

Index

344

Related titles published by Helion & Company

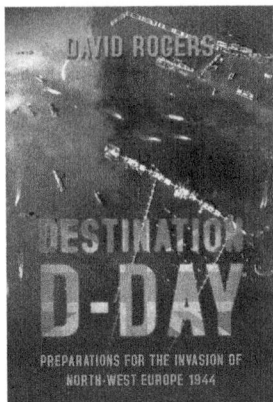

Destination D-Day. Preparations for
the Invasion of North-West Europe
1944
David Rogers
ISBN 978-1-909982-05-5
(Paperback)

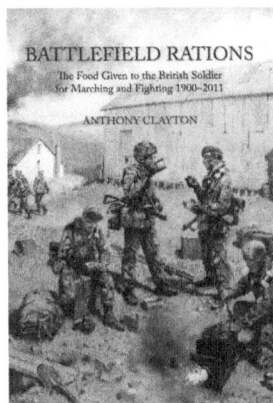

Battlefield Rations. The Food
given to the British Soldier
for Marching and Fighting
1900–2011
Anthony Clayton
ISBN 978-1-909384-18-7
(Paperback)

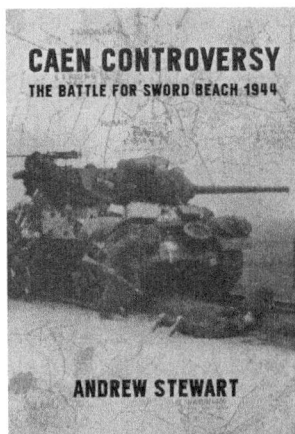

Caen Controversy. The Battle
for Sword Beach 1944
Andrew Stewart
ISBN 978-1-909982-12-3
(Hardback)

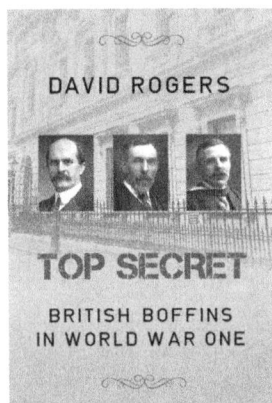

Top Secret. British Boffins
in World War One
David Rogers
ISBN 978-1-909384-21-7
(Paperback)

HELION & COMPANY
26 Willow Road, Solihull, West Midlands B91 1UE, England
Telephone 0121 705 3393 Fax 0121 711 4075
Website: http://www.helion.co.uk